SUBDIVIDING THE LAND

Metes and Bounds and Rectangular Survey Systems

SUBDIVIDING THE LAND

Metes and Bounds and Rectangular Survey Systems

Gaby M. Neunzert

CRC Press
Taylor & Francis Group
Boca Raton London New York

CRC Press is an imprint of the
Taylor & Francis Group, an **informa** business

CRC Press
Taylor & Francis Group
6000 Broken Sound Parkway NW, Suite 300
Boca Raton, FL 33487-2742

First issued in paperback 2019

© 2011 by Taylor and Francis Group, LLC
CRC Press is an imprint of Taylor & Francis Group, an Informa business

No claim to original U.S. Government works

ISBN-13: 978-1-4398-2747-5 (hbk)
ISBN-13: 978-0-367-86481-1 (pbk)

Library of Congress Cataloging-in-Publication Data

Neunzert, Gaby M.
  Subdividing the land : metes and bounds and rectangular survey systems / Gaby M.
Neunzert.
    p. cm.
  Includes bibliographical references and index.
  ISBN 978-1-4398-2747-5 (alk. paper)
  1. Surveying. I. Title.

TA545.N48 2010
526.9--dc22   2010016286

Visit the Taylor & Francis Web site at
http://www.taylorandfrancis.com

and the CRC Press Web site at
http://www.crcpress.com

Contents

About the Author

Gaby M. Neunzert came to this country from Switzerland in 1954 by way of Beirut, Lebanon. Following a 3-year side trip with the U.S. Army, Gaby graduated from the Colorado School of Mines in petroleum engineering. After graduation, a "missile" job, and an M.S. degree in petroleum engineering, Gaby was given the opportunity of a lifetime when in 1979 he was asked to start and also teach the surveying program at Red Rocks Community College. Ten years later came the call to teach surveying, and other topics in the civil engineering option at Mines. From 1991 until his retirement in 2000 as professor emeritus, he wrote an extensive GPS text, *Parting of the Land*, a user's guide to metes and bounds, as well as several surveying manuals. Gaby is a licensed surveyor in Colorado. For the near future, Gaby would like to share his knowledge of surveying, GPS, and general knowledge with as many people as possible. Since "retirement" Gaby has been actively promoting the Bill McComber Mentoring Program.

In Their Footsteps We Do Follow

This compilation of surveying information relating to the parting of the land is dedicated to three classes of surveyors, namely the Government Surveyors who subdivided "the big picture" of about two thirds of North America by dividing the land into one-square-mile sections, then the private surveyors who followed closely behind and, to this day, subdivide the land into ever smaller parcels which could be sold to individuals and, finally, unrelated to anybody else, are the geodetic surveyors who in the past mostly worked at night.

It probably is little realized by the public at large that, at least area-wise, the GLO (General Land Office) surveys are the largest and undoubtedly one of the most beneficial public works ever undertaken. As a modern person it is difficult to imagine that many thousands of individuals, deputized as Government Surveyors, would set out on foot into the wide-open spaces and survey either at $2/day or at $3/mile, whichever was cheaper for the government. As they worked, in groups of 6 or 8, they would start at a "known" point, align north or east, and measure a half mile, set a monument which in many cases has lasted better than 100 years, and repeat the process across the endless vastness of the forests, prairies, and many mountain ranges, from the bayous of Louisiana to the frozen tundra of Alaska. Most of the initial surveys were made from about 1830 to maybe the 1870s, but the process is still not yet complete. In Oregon the land was not opened up for settlement until the land was surveyed and Section 16 of each township was dedicated as the school section (640 acres).

Following the Government Surveyors was almost literally an army of private surveyors who started at the established monuments and then subdivided the land into 160 acres, then 80 acres or even 40-acre homestead land grants or into towns with 330 ft × 330 ft city blocks. First, a town was incorporated, then platted and finally surveyed on the ground into lots, which then could be sold to individuals. Many hours were spent by first making a drawing and then surveying out in the field untold miles of roads, from the size of a driveway to a freeway, and many more miles of railroads—on the plains, into the mountains, and then to many remote mining camps. While track-laying railroad crews had the reputation of being boisterous and hard drinking, surveyors would quietly come in small groups, set some stakes, and then disappear again. Utility lines had to be surveyed for electric companies, underground facilities such as the sewer, water, and natural gas lines and, yes, a private surveyor had to stake out the location of water reservoirs, river locks, canals, and irrigation ditches. Private surveyors have to be licensed in each state in which they practice, and since a property line can only be established through legal court action, a registered surveyor can only render an opinion for the location of the line based on measurements and appropriate calculations. Sometimes it can happen that the same surveyor gets praised by one landowner and blamed for the same work by an adjacent landowner.

Probably the least known to the general public are the geodetic survey-ors whose century-long task was to survey the nation with long triangu-lation chains and ultimately combine their data with geodetic surveyors from other nations for the size and shape of the whole earth. Until recently few people, other than geographers, appreciated their efforts, but with the U.S. Navy's need to have their submarines passively pinpoint their location at sea, GPS was postulated in the 1960s. All of a sudden known ground locations proved to be the key for determining the position of the GPS and later GLONASS satellites. Just as three lighthouses of known coordinates along the coast will fix the location of a ship, four satellites with known positions in the sky will give a fix for an observer on the ground. Maybe some of the better known geodetic surveyors are Charles Mason and Jeremiah Dixon of the Mason-Dixon line, who surveyed the border between Pennsylvania and Maryland from 1763 to 1767, and Sir George Everest for the survey of the Indian subcontinent during the first half of the 19th century.

Gaby M. Neunzert
Golden, Colorado

Historical Perspective

The roots of property surveys (cadastral surveys of ownership for the purpose of taxation) probably date back 4,000 years or more to Mesopotamia and Egypt, and have been culturally modified since that time. Thus, in order to understand the American property surveys, a brief historical overview is necessary.

The original inhabitants of North America did not have any written records of personal land property, and it was with the coming of the colonial conquests that cadastral surveys started. Beginning in the 1600s the Spaniards and later the Mexican Land Grant property deeds were made, and some are still valid, but only after adjudication by the U.S. judicial system. Since the 1650s the French left their legacy in the form of legal principles and "French Surveys." Both the French and Spanish based claims to the land on the right of discovery and conquest; the British on the other hand claimed the land through occupancy and use. "Naturally," British common law has applied to property in the original colonial states along the east coast since the 1600s.

Most of the early land was deeded rather slowly to a few individuals by local courts; in addition, large parcels of land, and with vague descriptions because of ignorance, were deeded by European monarchs. For example William Penn (1644–1718) received 45,000 sq. mi. or about 28,800,000 acres from King Charles II to form the colony of Pennsylvania.

In December 1791 the Bill of Rights was ratified by the First Congress, and Amendment V states: "… nor be deprived of life, liberty or property without due process of law; nor shall private property be taken for public use without just compensation." Without dwelling on the justifications, this act must rate as a paradox in light of the subsequent huge land "acquisitions" by the U.S. Again later, on July 28, 1868, the 14th Amendment was ratified, which similarly states: "… nor shall any State deprive a person of life, liberty, or property, without due process of law … with Congress to have the power to enforce, with appropriate legislations the provisions of the article." In passing it should be noted that land such as roads, etc., cannot be "condemned for the public good" without just compensation and due process of law.

In 1803, the United States roughly doubled in size with the Louisiana Purchase from France, with the land described generally: from the Mississippi River west to the Continental Divide and from the Mississippi Delta north to the 49th parallel, the present U.S.–Canadian border. Since there were no maps available and Lewis and Clark had not yet made their expedition westward, the signatories only knew that the Continental Divide was somewhere between St. Louis and San Francisco, and consequently were probably unaware that they signed for the area plus or minus many thousands of square miles. An editorial note: At 3 cents/acre, the Louisiana Purchase was probably the largest bargain in the history of the world, described by a metes and bounds deed. Other land acquisitions by the United States followed (see the map and table in the Surveying Roots section).

By contrast, individual landownership followed the colonists by different routes. The Spaniards primarily colonized Florida, New Mexico, Arizona, southern California, and southern Colorado. They granted comparatively few, but rather large, land grants to individuals along mostly the east coast of Florida and then westward into what is now Texas. Later, the Mexicans made "hundreds" of land grants to individuals from Texas westward to California. Ultimately, by about 1900, only roughly 10% of the original grants were adjudicated by U.S. courts. The French, on the other hand, followed the path of least resistance and the trade routes, mostly along coastlines (Great Lakes, etc.) and rivers (Mississippi and St. Lawrence, etc.), among other places. Private citizens laid out French lots *in aprents* "perpendicular" to the shoreline, many of which are still visible from satellites to this day. The vast majority of land prior to about 1850 was deeded by metes and bounds, initially under British and later American law. Land was first staked out, then surveyed and ultimately recorded in the city or county. The surveys started at some arbitrary point (for example Pilgrim Rock), mostly along the coast, and one survey after another was tied to the previous one in a quilt-like, typical metes-and-bounds pattern. Over time, land ownership slowly progressed westward to the Appalachian Mountains.

The U.S. Rectangular Survey System was initiated by Congress in 1785, and the first surveys were made in Ohio between 1786 and 1800. The population pressures to acquire private lands and the need to sell land in order to finance public projects was comparatively small. This all changed dramatically with the large migration of people over the Oregon Trail into the Willamette Valley of Oregon (1840 to 1850), the California Gold Rush (1849 to 1860), and the completion of the transcontinental railroad in 1869. People buying land and moving goods in an orderly fashion were needed to fill the vast spaces of the continent, and to the great envy of other large nations, the U.S. Land Survey System made this possible. Even at $1.50/acre, large amounts of money could be raised by governments and railroads by selling land which had been surveyed. Schools could be financed by selling land in Section 16 and Section 36. Roads could be built along section lines, and city blocks could be laid out in an orderly fashion by subdividing a section. Thus, the stage for the property surveys in the United States was set.

Introduction

The great American dream of owning a house and surrounding land has to be looked at carefully, since it probably represents an individual's greatest lifetime financial investment and absolute ownership is not always guaranteed. Through their lending practices, banks probably "own" more real estate than the individuals making the mortgage payments; on the other hand, few individuals realize that ownership involves "proper paper documentation," as well as "complete physical documentation," and one without the other could spell expensive litigation. Paper documentation involves the necessary and complete legal descriptions and proper filing with the courts, and the physical documentation requires surveying type measurements and calculations to locate property corners, correct alignment of fences, etc. Complicating matters is the fact that there are no federal or global "rules," and every state and sometimes even just local jurisdictions have their own individual surveying laws and regulations. Understanding some of the "roots" can be very helpful. Initially, a property owner is probably only concerned with the size, in acres or square feet, of the property purchased, but one quickly has to learn about how real estate is described when a dispute about the boundary arises. Paraphrasing poet Robert Frost that "good and visible fences make good neighbors" has a corollary that "no fences (or an undetermined boundary) can make bitter enemies of adjacent neighbors."

This text, entitled *Subdividing the Land—Metes and Bounds and Rectangular Survey Systems*, is a basic compilation of the Metes and Bounds concept and the U.S. Rectangular Land Survey System. It was written for anybody who is or will be in contact with any real estate transaction—surveyors, realtors, lawyers dealing with real estate, and probably most importantly, all individuals who will take out the biggest loan of a lifetime for a dream house and property.

The main body of material is dedicated to the two basic concepts—metes and bounds, and the U.S. Rectangular Survey System as an overview across the entire U.S. Care should be taken to note that "metes and bounds" only controls the land boundaries essentially in the 13 original colony states along the eastern seaboard, as well as Texas and Hawaii (Figure 1). In the remainder of the U.S., even with a modern deed, the survey has to start at a GLO monument as the primary survey and then enclose the property with a metes and bounds description as the secondary survey (Figure 2). No purchase of land is ever complete without the necessary legal wording, and for this purpose some basic explanations are given in the section, "Some Legal Concepts and Definitions." Frequently, the water flowing through a property or the groundwater and the mineral rights are independent and are not necessarily

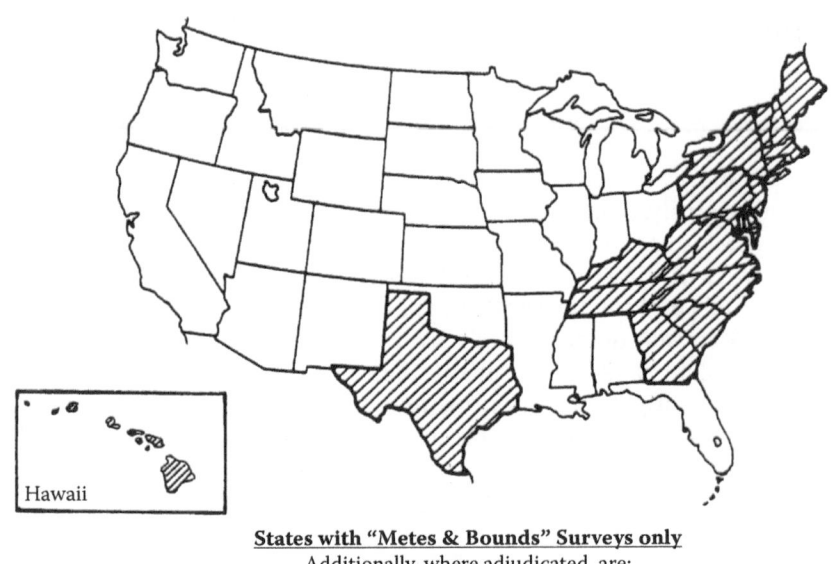

States with "Metes & Bounds" Surveys only
Additionally, where adjudicated, are:
French surveys,
Spanish & Mexican surveys
Ahupuaa surveys in Hawaii

FIGURE 1
States with "Metes and Bounds" Surveys only.

purchased with the surface of the land. Thus before looking seriously at property in the mineral belts of the western United States, the section on "Mining Claims and Related Items" should be consulted; also, since water is frequently more valuable than private and commercial land, reading the "Water Laws" section is recommended. Even though only partially adopted in the United States, the Torrens system should be investigated since it provides property registry by owner name, rather than the "usual" registry by metes and bounds description. Realistically, no surveying narrative would be complete without a historical section mentioning the roots of property law and the British, French, Mexican, and Spanish surveys, as well as the start of the GLO survey in Ohio.

Illustrations can often explain surveying concepts better than words or equations, therefore many original drawings are included throughout the narrative. Supplemental Comments precede the full page illustrations, Figure 25 to Figure 51. Surveying concepts are both global and regional, and the author generally has made an effort to present the material on a universal basis: however, some examples and illustrations are local (Colorado or even Denver) in order to make a specific point.

By now the reader should be well aware that the topics mentioned above will fill a sizable portion of a library and that the material on hand can at

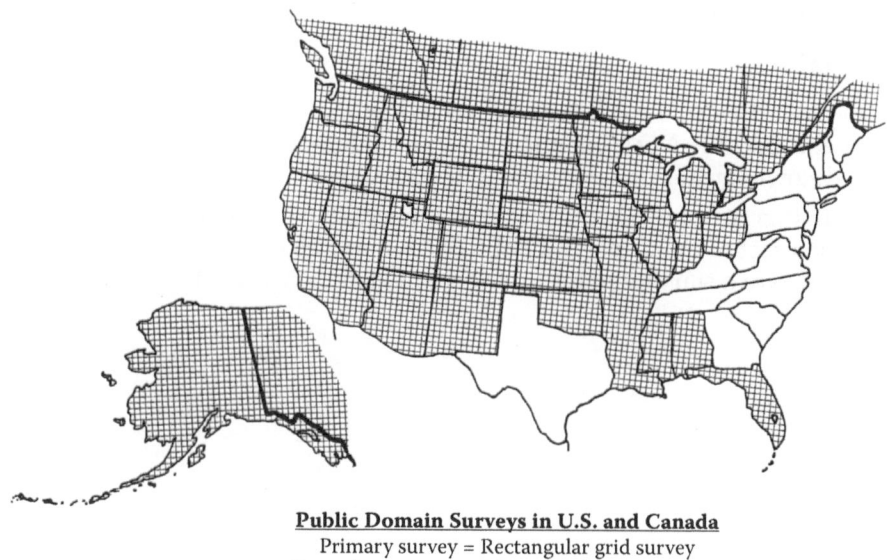

Public Domain Surveys in U.S. and Canada
Primary survey = Rectangular grid survey
Secondary surveys = Metes and bounds surveys

FIGURE 2
Public Domain Surveys in U.S. and Canada.

best be an abstract, which is only the beginning. Initially the component parts for *Subdividing the Land* were written for novices as a practical introduction for basic classes and workshops because the topics were not or only partially covered in the available literature. The concept of introductions still underlies this presentation but over the years many gems of wisdom have been added, which can make the material a valuable reference for even the professional. For example, for the "practical" individual are the sections of what constitutes a monument, what evidence does one have to look for, and what physically was "set" in the ground; all topics which are usually not covered in theoretical, conventional textbooks. For the self-learner there may be an initial difficulty with the "vernacular" considering that the vocabulary in many parts is derived from a technical/surveying/civil engineering as well as from a legal background. Reading on though the paragraph and in context should clear up most jargon; to start with, a "glossary" is included as the last chapter, otherwise a comprehensive dictionary can help. Unlike most other topics relating to surveying, there is no math associated with the "legal" aspect of property surveys. It is also true that this material is written from a surveying perspective, where theory and applied real life facts meet.

Finally, it is very strongly advised that one or more licensed professionals in the applicable state be consulted before undertaking any real transaction, since there are only state, and no general federal property laws.

Disclaimers

a) Even though the concepts presented below are based on solid laws, this book is NOT intended as a law reference on real property. A knowledgeable lawyer should be consulted.

b) In this age of "everything you ever needed to know," this paper is an abstract of the principles only. The references listed in the bibliography are a starting point for the many details involved.

Metes and Bounds Surveys

Metes = to measure
Bounds = the boundary of the land

Not just in the United States but all over the world, wherever private property rights are respected, a metes and bounds survey description is the basic type of property deed. In its most elementary form, a metes and bounds description must:

Start at a stated point of beginning (P.O.B.), thence describe each leg by direction and length, and ultimately return to the point of beginning.

There are many additional aspects that are either directly stated or implied.

Direction of Each Leg

True versus Magnetic Directions

- Prior to about 1900, most directions were obtained by magnetic compass and are thus subject to magnetic declination (also called *magnetic variation*), which probably is seldom stated.
- The magnetic declination is the deviation from true north and is subject to annual changes. Within the continental United States, magnetic declinations of up to 25° either east or west are possible, with annual changes of a half degree per year. In addition, there are areas with very large magnetic anomalies, for example, in the Iron Ranges of northern Minnesota.
- Since about 1900, most directions are stated by implication from true north and are ideally based on an astronomical observation.
- Since about 1995, directions can also be obtained by the Global Positioning System (GPS), which requires that the mapping angle is either stated or was included in the calculations for the direction.

Type of Direction

- The type of direction implies a direction of travel around the property, either clockwise or counterclockwise.

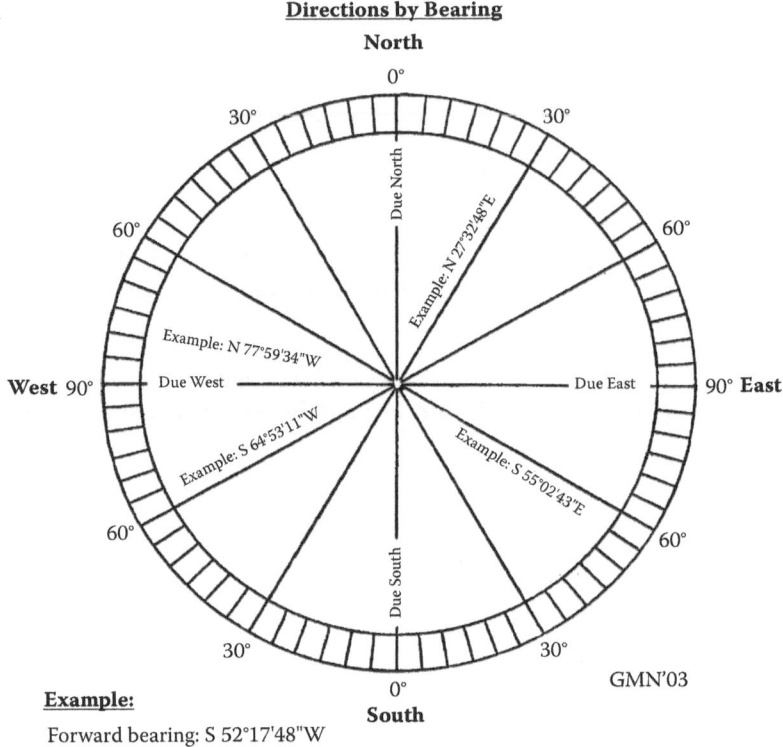

Directions by bearing.

Example:
Forward bearing: S 52°17'48"W
Reverse bearing: N 52°17'48"E

Max. angle = 89°59'59"

FIGURE 3
Directions by bearing.

- American usage favors bearings such as Due North, Due East, Due South, or Due West, or the quadrant designation of, for example, N 31°47'22" E (Figure 3).
- Non-American usage favors azimuth north as a direction. An azimuth is defined as a clockwise (angle right) angle from true north. Starting with due north = 0° Az, due east = 90° Az, etc. For example, a bearing of N 31°47'22" E becomes an Az = 158°12'38". Computer usage will probably encourage the use of azimuth in this country (Figure 4).

Distances of Each Leg

Unless otherwise stated, all distances are reported as horizontal distances; slope distances are either mechanically or mathematically reduced to horizontal.

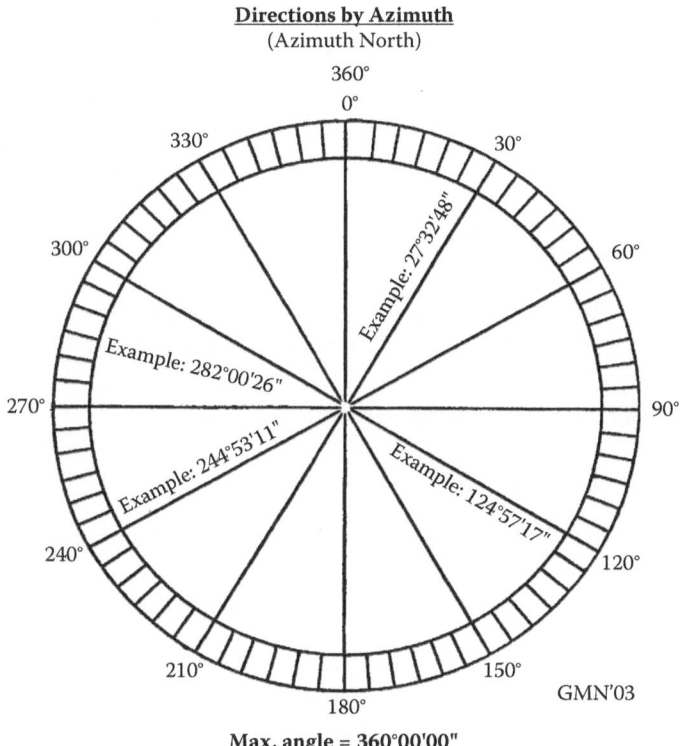

Directions by Azimuth
(Azimuth North)

Max. angle = 360°00'00"

FIGURE 4
Directions by azimuth.

Distances less than 1 mile (with GPS, less than 1/2 mile) in length are considered to be on a flat plane, longer lines follow the curvature of the earth or an ellipsoid (mathematical model).

Units of Length

- The basic unit of surveying distance is the U.S. Survey foot, which is usually subdivided into 1/10 ft. and 1/100 ft. Distances are normally reported to the nearest 1/100 ft. 1 U.S. survey foot = 1.0000020 U.S. standard foot. See the explanation of "Foot" in Glossary.

- The metric unit of distance is the international meter;
  39.37 in. = 1.000 m.

- By statute, the basic unit of the U.S. Land Survey System is the Gunter chain; 1 Gunter chain = 100 links = 66.00 ft; one link = 0.66 ft.

- The *vara* was the original unit for many Spanish surveys; it varies in length and is about 33 inches. Consult a local surveyor for the specific value used.
- The *arpent* was the original unit for French surveys; technically, it is an area measure; 1 arpent = 0.85 acres, the sides of a square arpent = 192 ft; it varies in length.

And then there were:

- Distances determined by counting the number of revolutions on a wagon wheel with a flag tied to a spoke; most of the time the size of the wheel is unknown.
- Distances determined as "a half hour ride," etc.

Determination of Distances

The default method for distances is: a length measured with a steel (or invar) measuring tape.

There are two types of surveys that must be considered:

1. New survey, i.e., a survey over ground that has not been surveyed before
2. Resurvey, i.e., a survey that follows in the footsteps of the previous surveyor

New Surveys

A new or initial survey is made when the land is first subdivided. The table below is intended as a guide. In general, the more expensive the property, the higher the survey ratio has to be. Implied in the table is a mathematical check, i.e., a closed traverse, and not some free-standing "side shots" (see Glossary for definitions).

Method	Error-to-distance ratio	Application
Pacing	1:100	Excellent for quick check in field
Ordinary taping	1:5,000 to 10,000	For low-cost property surveys
Precision taping	1:20,000 to 1:50,000	Modern property surveys
EDM (total station)	1:15,000 to 1:40,000	Modern property surveys
GPS	up to 1:100,000 possible	Selected modern surveys

CAUTION:
[a.] GPS field work MUST provide 2 independent sets of data for each property corner, i.e., single occupancy of each corner is NOT acceptable.
[b.] GPS calculations must consider ellipsoid used and elevation datum.
[c.] The final values must show the data base used, i.e., the ellipsoid (Clarke 1866, WGS'80, etc.), NAD'27, or NAD'83, etc. (see below for more details).

Resurveys

Once an existing property line has been established with two authentic and adjacent monuments, the resurvey must *"follow in the footsteps of the original surveyor"*; modern or better surveying methods will <u>only improve the accuracy, but they cannot fault an old survey.</u>

Logically and legally, it only makes sense to accept the monuments "of call" and not set a new monument(s) nearby. A monument "of call" or "of record" is a surveying monument described in the original or resurvey notes. As part of a resurvey, it is possible and indeed recommended that a defective monument be rejuvenated or even replaced with a new monument and, in turn, the changes are described in the field notes. The first rule for making a resurvey of a property is:

> ### The monument of record controls—both for distance and direction
> *If nature moves a monument, the property moves with it.*
> *If a human moves a monument, it is an illegal act.*

The fundamental question confronting both the novice and experienced surveyor is "What is, or was, the monument of record?" The answers can range from a shiny brass cap glistening in the sun, to "no clue" because the field notes merely state "set monument." The frustrations are only compounded when several monuments can be found in the immediate vicinity (i.e., the pin cushion principle), and all were set by some uninformed surveyors.

The second rule of a resurvey is:

> *A diligent search for and subsequent locating of a monument of record can answer many questions.*

Additional Aspects of Metes and Bounds Descriptions

- In the "metes and bounds states," (see Figure 1) starting at the point of beginning (POB) is sufficient. In the "GLO [General Land Office] states," (see Figure 2) the point of beginning must be tied to a public monument, such as a section corner, etc.

- Probably the most valuable addition to the basic metes and bounds description is the description of the monuments (and possibly ties) set during the original survey.

- *Area.* Most property is not sold by a boundary description but rather by area; as in "containing 2.54 acres more or less." As stated above, distance and areas are not measured on the slope but are reduced to the horizontal.

- An indirect metes and bounds description with reference to a map or plat is possible. For example: Lot 3, Block 15, First Filing, Rolling Hills Subdivision, etc., or Unit 26, Building 5, Green Meadows condominium, etc. The map or plat in turn should give the necessary boundary description (i.e., distance and direction).

- It is not necessary that all boundaries are described by straight lines; a property following a horizontal curve of a road, and the curve (or a portion of it), can be described by radius, deflection angle, etc. Boundaries along a river or lake can also be described by straight line segments and arcs. Condominiums require a three-dimensional description.

- Even though modern metes and bounds descriptions are mathematically sound and rely on "man-made" monuments, older descriptions often rely on "calls" to natural objects, such as "To the fork (branch) of the Rouge River, thence to an Aspen tree with a six-inch blaze on the north side," etc.

Following below are three examples of metes and bounds description, arranged in increasing complexity:

1. A written description only, from North Carolina with references to natural monuments
2. A written description and plat (see Figure 5) of a survey in a GLO state
3. A written description and plat (see Figure 6) of a 25-acre parcel in a GLO state

Example #1—Metes and bounds from North Carolina, with references to natural monuments, is for an 8.4-acre tract at \$125/acre with the final payment due August 18, 2015.

Tract One is described as follows:

> *Beginning at a stake, wild cherry pointer, W.S. Gateway's corner, running thence South 31° East, 7.75 chains to the center of the public road; thence as the road follows, viz. South 72° West, 11 chains to the bridge at Red Branch; thence North 57° West, 1.5 chains; thence North 38 1/2 ° West, 4.25 chains to a stake by the public road, Howard Marsh's corner; thence this line North 58° East, 13 chains to the point of beginning, containing 8.4 acres, more or less.*

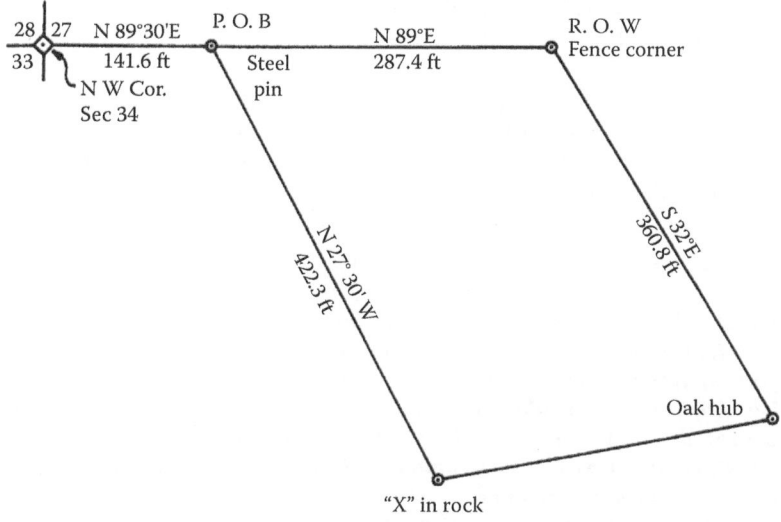

FIGURE 5
Example of simple metes and bounds survey.

Example #2—Simple metes and bounds and plat of a survey in a GLO state (Colorado)

There are several necessary comments:

1. The description must be in the "Western U.S." since the tie is made to the GLO System for the primary surveying control.
2. The plat, originally made from the written description, is shown intentionally to be "inferior" because the bearing and length of the leg from the "oak hub" to the "X in the rock" has been omitted. This can have some serious implications when the original description has been lost or destroyed. For more details, see section titled Some Legal Concepts and Definitions.
3. Judging from the angular values, i.e., to the nearest minute of arc and distances to the nearest 1/10 ft, this property was probably surveyed (with transit and chain) and the description written in the 1930s.

The property description and plat are shown below:

> *Beginning at the NW Corner, Sec.34, T3S, R70W, 6PM, and running N89°30'E, 141.6 ft to a steel pin, the true point of beginning. Thence, N 89°E, 287.4 ft to the R.O.W. fence corner; thence S 32°E, 360.8 ft to an oak hub; thence S 78°W, 290.8 ft to a cross chiseled on a rock; thence N 27°30'W, 422.3 ft to the true beginning and containing 3.17 acres more or less.*

Example #3—Metes and bounds and plat of a 25-acre property in a GLO state, after H.A. Babcock, 1960

Shown below is an example of a metes and bounds description for the purpose of showing some possible variations: different monuments, straight, curved and meandering boundaries, etc. (Obviously, some names have been changed to prevent any reference to an existing survey.)

The property is described and plat are as follows:

A parcel of land in Green County, Pennsylfornia, lying wholly within the South West 1/4, of the South East 1/4, of Section 10, Township 3 South, Range 53 West, Coyote Meridian, surveyed by Bob Plumb, and more particularly described below. Any reference made to corners or fraction of a section, is in Section 10, as shown above. All bearings refer to the true meridian, as obtained by solar observation on Course No. 1. All monuments were in position at the completion of the survey on April 15, 1954.

Beginning at the S 1/4 corner, Section 10, T3S, R53E, Coyote Meridian, the true point of beginning, marked by a standard U.S. Bureau of Land Management brass cap.

1. *Thence, N 89°23'E, 593.70 ft, along the south boundary of the section, to a 3 in. range box, with a brass cap marked RLS 1327, which marks the westerly right-of-way of U.S. Highway 827.*
2. *Thence, following the westerly R-O-W, N 15°20'E, 426.80 ft to point of curvature of a horizontal curve, marked with a 6 in. x 12 in. concrete Pennsylfornia R.O.W. monument.*
3. *Thence, from the tangent N 15°20'E on a curve to the right, radius 1507.39 ft and arc length of 1127.68 ft, to a #5 rebar, 2 ft long, driven within 3 in. of the ground, the end of the R.O.W.*
4. *Thence, following the NS centerline of the SE 1/4 of the section, N 10°34'W, 138.62 ft, to a 2 in. steel shaft set in concrete and marked on top SE 1/16 S 10, 1949, and on the east side HAB 1084.*
5. *Thence, following the EW centerline of the SE 1/4 of the section, S 89°32'W, 695.23 ft, to a car axle driven (1 1/2 in. dia.) about 2 1/2 ft into the ground, set as nearly as practical on the EW centerline of the SE 1/4 of the section and the mean high-water line on the south bank of the Muddy River.*
6. *Thence, following the meandering of the mean high-water line on the south bank of the Muddy River upriver, S 64°20'W, 463.79 ft to a 3/4 in. diameter steel pin set in a 6 in. diameter cylinder.*
7. *Thence, due west, 132.13 ft to a tack set in a 2 in. × 2 × in × 12 in. white oak hub, driven flush with the ground.*
8. *Thence, S 42°15'W, 247.72 ft to a 38 in. × 47 in. × 22 in. block of sandstone, buried on edge, with a 2 in. cross chiseled into it, set on the mean high-water line of Muddy Creek and the NS centerline of the section.*
9. *Thence, following the NS centerline of the section, S 20°43'E, 1050.36 ft, to the point of beginning.*

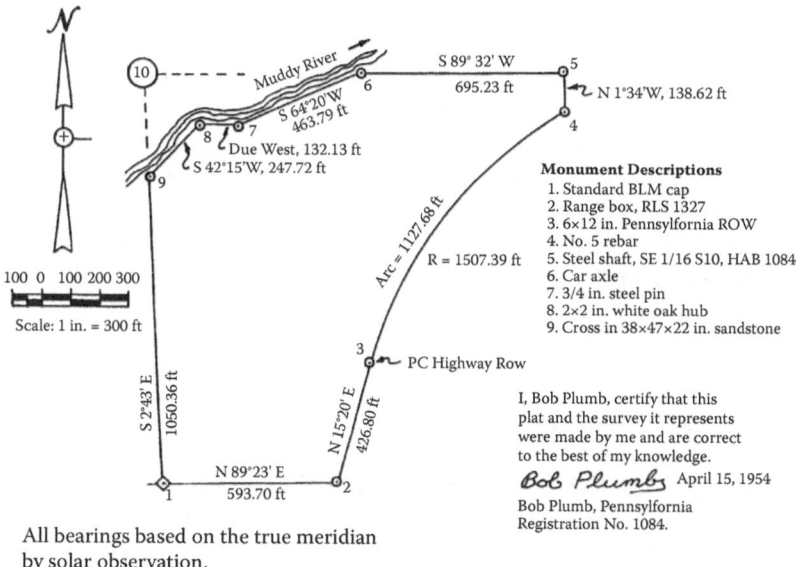

All bearings based on the true meridian
by solar observation.

FIGURE 6
Plat of 25-acre parcel in GLO state.

Metes and Bounds by Coordinates—The Future

Undoubtedly, the day will come, thanks to GPS and GIS databases, when property will be described by X and Y plane rectangular coordinates and for condominiums in 3-D, X,Y, and Z coordinates. By knowing the coordinates of each inflection point, it is very easy to calculate the distance and direction of each leg.

The concept is excellent; the execution, however, has some pitfalls.

From a technical standpoint, there are two possibilities for the initial surveys:

1. *A conventional survey* with angles and distances in the form of a closed traverse and later conversion to coordinates with a tie to a control point with known coordinates. This method is perfectly acceptable because it has a built-in *math check* and a *diagnostic closure ratio*.

2. *By obtaining GPS data at each inflection point.* As attractive as this idea is, it is *not* recommended because it offers *no math checks* and *no closure ratio* on the quality of the survey. Since there are no checks, RTK (Real Time Kinematic) GPS observations are *not* recommended, but there are, however, several other methods available that can be used for property surveys, since they provide a check on GPS data.

In line with the modern (computer) trend of working with plane coordinates, the calculations and ultimately the reporting and filing of the coordinate data require **close attention to details:**

a. The transformation from the "curvature" of the earth to the "flat coordinate plane," or back, is usually made with either the Mercator or the Lambert conformal projection, and

b. The ellipsoid; with NAD'27 the Clarke 1866 ellipsoid is used and for NAD'83 and later data bases the WGS'84 (GRS'80) ellipsoid applies, and

c. Various zones of "State Plane Coordinates" or UTM coordinates and finally

d. The "correct" data base

Without dwelling on the details, in Colorado, for example, there are by now many different data bases and for surveying accuracy it is very difficult to convert from one base to another, even though there are computer programs that will reputedly perform the task.

What Is the Data Base?

There are many answers:

a. Local coordinates. The least sophisticated concept, but probably used in most construction projects. Local coordinates normally apply to a very small area (usually less than 1 sq. mi.), it is assumed that the earth is flat, and the coordinates are at ground elevation (orthometric). Unless special provisions are made, local systems are completely independent and are not tied to the "world" (i.e., state plane coordinates, etc.), and, thus, when the project is finished all numbers usually become meaningless.

b. NAD'27 (North American Datum of 1927). The data base for the "original" State Plane Coordinate Systems; used until roughly the 1980s. No longer supported by NGS (National Geodetic Service); still shown on most USGS (U.S. Geological Survey) standard topographic 7½ᵐ maps.

c. NAD'83 (North American Datum of 1983). A data base established from an extensive readjustment of control stations throughout North America. No longer supported by NGS; shown only on the most recently published USGS maps. Since the adjustment of the stations was non-linear, there is <u>no single multiplier</u> for the conversion from NAD'27 to NAD'83 and even though the monuments have not moved, the numerical difference between NAD'27 and NAD'83 can be in the <u>order of *several hundred feet*</u>.

d. HARN'92 (High Accuracy Reference Network; Colorado 1992). A refinement of the NAD'83 data. No longer supported by NGS; shown only on the most recently published USGS standard 7½ᵐ topographic maps. Since the adjustment of the stations was non-linear, there is <u>no single multiplier</u> for the conversion from NAD'83 to HARN'92. The difference between NAD'83 and HARN'92 is in the <u>order of *several tenths of a foot*</u>.

e. NSRS'06 (National Spatial Reference System). The "latest" update of the national data base, effective October 11, 2006. Supported by NGS. The readjustment from HARN'92 to NSRS'06 was non-linear, and there is <u>*no single multiplier*</u> for the conversion. The difference between HARN'92 and NSRS'06 is in the order of <u>*a tenth of a foot* or less</u>.

f. The future. There undoubtedly will be other, more "modern" data bases.

g. **The bottom line**. As long as the **basis of the data base is stated,** any system is acceptable. Especially if the original surveying data is available, it is possible, with proper precautions, to convert from one system to another. **With GPS data it is also very important that the calculation parameters** (ellipsoid, type of projection, geoid elevation, etc.) **are preserved together with the published coordinate values.**

U.S. Rectangular Survey System

Laying Out the "Grid"

On March 20, 1785, the Continental Congress passed "an ordinance for ascertaining the mode of locating and disposing of lands in the western territory, and for other purposes therein mentioned"; thus the U.S. Rectangular Survey System was inaugurated. This same system is also known as the U.S. Public Land Survey System (PLSS) or the GLO Survey (Government Land Office Survey).

Together with the Canadian system, which was started about 1789, an estimated 11% of the earth's surface is covered by a rectangular survey system (see Figure 2 in the Introduction). On paper, the idea of dividing the country into orderly squares had, and still has, great appeal, but it suffers both logically and physically when the concept is applied to the "round" or spherical earth. Despite the conceptual shortcomings, the rectangular system provided an "instant" primary surveying data base for exploration and settlement, and allowed for an orderly growth that is unparalleled in history. As time progressed and land was further subdivided, metes and bounds surveys were then overlaid as a secondary survey and tied to the rectangular system (see Figure 2 in the Introduction).

At this point the reader should note that the presentation following below will differ markedly from the majority of surveying textbooks and other published material. Conventional explanations of the rectangular survey system often leave the impression that the GLO states are neatly covered by a myriad of perfect squares. Wrong, for two reasons:

1. It is impossible to fit a plane square pattern over the curved surface of the earth, and

2. Even under perfect conditions genuine surveying errors will occur and provisions have to be made to account for them. As to be expected in the 1800s and early 1900s, conditions were far from perfect and, yes, there were even some fraudulent "barroom surveys."

The explanation following will attempt to combine the paper concept with actual field surveying results.

Another word of caution: Surveying reflects the character of its time, and this is no different with government surveys, which were controlled by *The Manual of Surveying Instructions*, published in 1855, 1881, 1890, 1894, 1902, 1919, 1930, 1947, and 1973, and the newest edition of 2009. Preceding the manuals are instructions to Deputy Surveyors in the state of Mississippi (1831), Territory of Arkansas (1837), the States of Illinois and Missouri (1834), Territory of Florida (1842), Territories of Wisconsin and Iowa (1846), Ohio, Indiana, and Michigan (1850), and the public lands in Oregon (1851). The principles below are generalized and for the specific instructions in a local area the reader is referred to the manual. The manual to consult is the manual in force at the time of the survey (i.e., it had to be published before the survey). Reading the latest edition will not help to decipher what was done over 100 years ago!

In this text, the term *error* requires an explanation, because it is used in a surveying context. It is a "true fact" that every physical measurement, including all surveying measurements, contains an error, and the corollary is that no measurement is ever exact. Over the years, the allowable or acceptable error(s) in surveying have decreased, but nonetheless errors are a fact of life to be taken into account and not a fault which can be avoided.

With a multitude of measurements, there are both physical and mathematical procedures that must be utilized to distribute (adjust) errors, thus further modifying the original measurement. In the review of previous surveys it is, even for the professional, often rather difficult to determine the source of errors, whether it was the original equipment or methods, or subsequent mathematical adjustments.

Even though, strictly speaking, it is not an error, but with the advent of computers and GPS, the nature of the data bases has become a major problem (for example, NAD'27, NAD'83, NSRS'06, etc.). As a statement, without explanation, it is now possible to make a "perfect" measurement on the ground and have this value appear appreciably different in coordinates from the original and yet different again from one data base to the next.

Historical perspective aside, modern surveyors and other users have to be aware of the reasons and methods of the original GLO surveys, in order to understand the public land monument patterns, perform resurveys, and interpret legal conflicts (Figure 7).

The Principal Points

For the rectangular survey system there are 37 starting points, called *principal points*, plus the original surveys in Ohio (Figure 7 and Figure 8). In passing, it would follow that if a deed does not make specific reference to a principal point, a property could be described in 37 different locations. From each of those principal points, surveys were run in all directions until another government survey was encountered, and to this day there are still some federal lands which have not been formally surveyed. The coverage varies appreciably; for example, as shown in Figure 7, the area covered by the 5th Principal Meridian stretches from the Gulf coast to the Canadian border, but is comparatively narrow in an east–west direction; the area covered by the 6th Principal Meridian is basically east–west; the Ute Principal Meridian covers only a few square miles around Grand Junction, Colorado, etc.

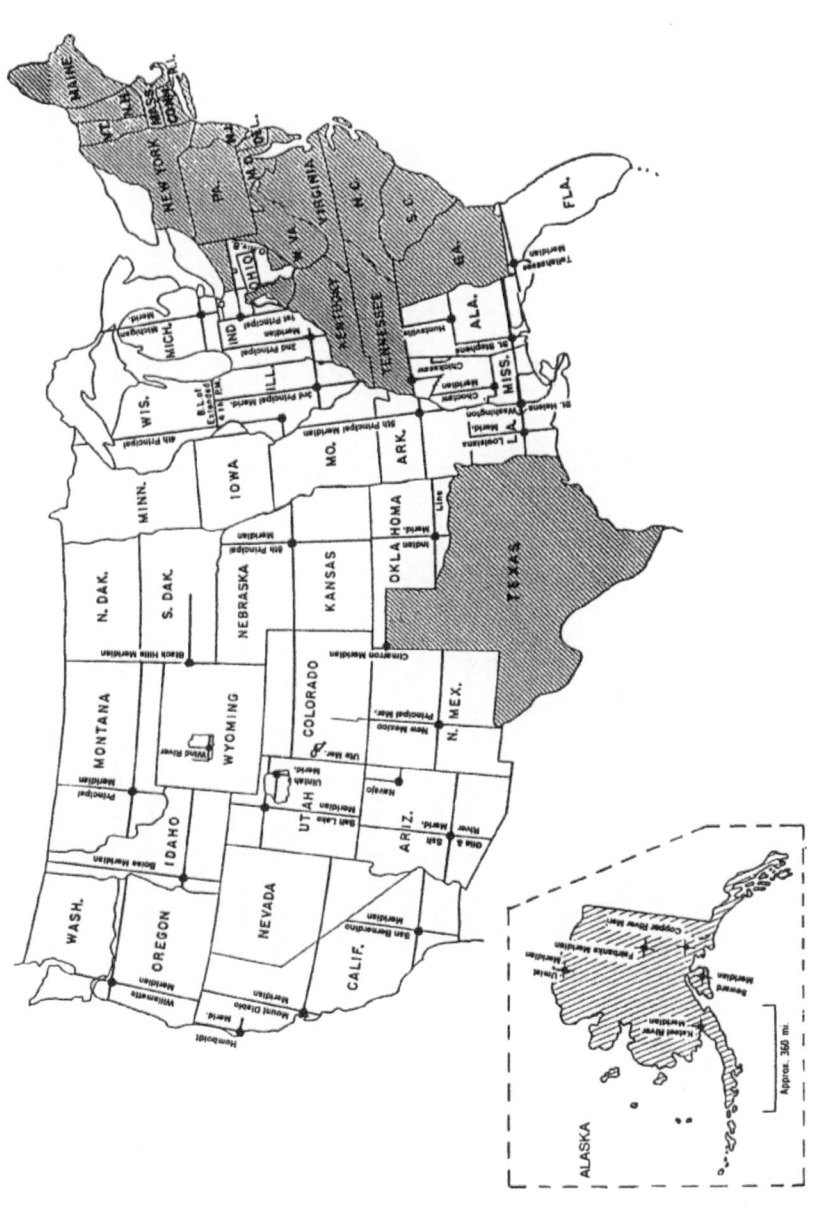

FIGURE 7
Principal Meridians in the U.S. rectangular system.

Meridians and base lines of the United States rectangular surveys

Meridian	Adopted	Governing surveys wholly or in part in states of	Initial points					
			Latitude			Longitude		
			°	′	″	°	′	″
Black Hills	1878	South Dakota	43	59	44	104	03	16
Boise	1867	Idaho	43	22	21	116	23	35
Chickasaw	1888	Mississippi	25	01	58	89	14	47
Choctaw	1821	Mississippi	31	52	32	90	14	41
Cimarron	1881	Oklahoma	36	30	05	103	00	07
Copper River	1905	Alaska	61	49	04	145	18	37
Fairbanks (1)	1910	Alaska	64	51	50.048	147	38	25.949
Fifth Principal	1815	Arkansas, Iowa, Minnesota, Missouri, North Dakota, and South Dakota	34	38	45	91	03	07
First Principal	1819	Ohio and Indiana	40	59	22	84	48	11
Fourth Principal	1815	Illinois	40	00	50	90	27	11
Fourth Principal	1831	Minnesota and Wisconsin	42	30	27	90	25	37
Gila and Salt River	1865	Arizona	33	22	38	112	18	19
Humboldt	1853	California	40	25	02	124	07	10
Huntsville	1807	Alabama and Mississippi	34	59	27	86	34	16
Indian	1870	Oklahoma	34	29	32	97	14	49
Kateel River (2)	1956	Alaska	65	26	16.374	158	45	31.014
Louisiana	1807	Louisiana	31	00	31	92	24	55
Michigan	1815	Michigan and Ohio	42	25	28	84	21	53
Mount Diablo	1851	California and Nevada	37	52	64	121	54	47
Navajo	1869	Arizona	35	44	56	108	31	59
New Mexico Principal	1855	Colorado and New Mexico	34	15	35	106	53	12

			Latitude			Longitude		
Principal	1867	Montana	45	47	13	111	39	33
Salt Lake	1855	Utah	40	46	11	111	53	27
San Bernardino	1852	California	34	07	13	116	55	48
Second Principal	1805	Illinois and Indiana	38	28	14	86	27	21
Seward	1911	Alaska	60	07	37	149	21	26
Sixth Principal	1855	Colorado, Kansas, Nebraska, South Dakota, and Wyoming	40	00	07	97	22	08
St. Helena	1819	Louisiana	30	59	56	91	09	36
St. Stephens	1805	Alabama and Mississippi	30	59	51	88	01	20
Tallahassee	1824	Florida and Alabama	30	26	03	84	16	38
Third Principal	1805	Illinois	38	28	27	89	08	54
Uintah	1875	Utah	40	25	59	109	56	06
Umiat (3)	1956	Alaska	69	23	29.654	152	00	04.551
Ute	1880	Colorado	39	06	23	108	31	59
Washington	1803	Mississippi	30	59	56	91	09	36
Willamette	1851	Oregon and Washington	45	31	11	122	44	34
Wind River	1875	Wyoming	43	00	41	108	48	40

(1) U.S.C. & G.S. station "Initial, 1941" is located S. 66° 44' E., 2.85 feet distant from the initial point of the Fairbanks meridian. The geodetic station (latitude 64° 51' 50.037" N., longitude 147° 38' 25.888" W.) was inadvertently used as the origin from which to compute positions on the Fairbanks Meridian protraction diagrams.

(2) The Kateel River initial point is identical with U.S.C. & G.S. station "Jay, 1953."

(3) The Umiat initial point is identical with U.S.C. & G.S. station "Umiat, 1953." Positions are as published by the United States Coast and Geodetic Survey.

FIGURE 8
Table of principal points.

After establishing the "initial" or principal points as needed, the federal surveyors of the public lands surveyed either true north or true south, called the Principal Meridian, and set monuments at 1 mile (80 chains) or ½ mile (40 chains) intervals. From the principal point, the Baseline was laid out, due east or due west, along a small circle route and monuments (called standard corners) were set at 1 mile (80 chains) intervals (see Figure 11). Depending upon the location of the line in the survey pattern and the time period, the direction of the meridian lines was either established with an astronomical observation (solar or polaris) or by magnetic compass.

Laying Out the Quadrangles

But with the congressional mandate to divide the land into "squares," two thoughts had to be kept in mind: one, the distribution of errors, which certainly are involved, and, two, the curvature of the round earth. By following a very specific surveying sequence and first dividing the land into 24 × 24 mile blocks, called **quadrangles** or tracts, and then into 6 × 6 mile **townships**, the errors could be distributed to the north and east (see Figure 12 and Figure 13). Later, with an additional subdivision into 1 × 1 mile blocks, called **sections**, gross surveying errors could be redistributed to the north and west (see Figure 15 and Figure 16).

Laid out on the round earth, a fundamental logic difficulty arises when a "square" is defined by meridians and parallels. Meridians are north–south lines which converge at the pole and parallels are east–west lines (lines of equal latitude, also called small circles). It may be difficult to see on the sketch, but the meridian north–south line is a straight line, whereas an east–west line is an arc (small circle), with its radius depending on the latitude. As shown in Figure 9, a 24 mile "square," because of convergence, has to be less than 24 miles on its north side, even though the other three sides can be exactly 24 miles. The illustration and the numerical example below also show that the convergence becomes more pronounced with increasing latitude.

Location	Approximate latitude	Northern Boundary of 24 mile "square" shorter by
New Orleans, Louisiana	30° N	446 ft
Boulder, Colorado	40° N	647 ft
49th parallel (U.S.–Canada line)	49° N	887 ft
Point Barrow, Alaska	72° N	2,374 ft

Back to the quadrangles of 24 × 24 mile squares. The basic figure (Figure 9) was redrawn into Figure 10, to show that the full 24 mile width had to be surveyed every 24 mile interval north or south at the standard parallels or correction lines, in order to offset for convergence. Also see Figure 11 and Figure 14. Just to illustrate the merit of reading the manual applicable at the time of survey, the reader is referred to Figure 27 showing the Denver,

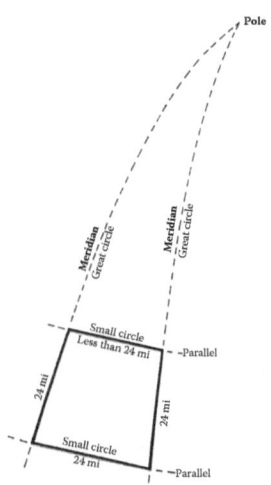

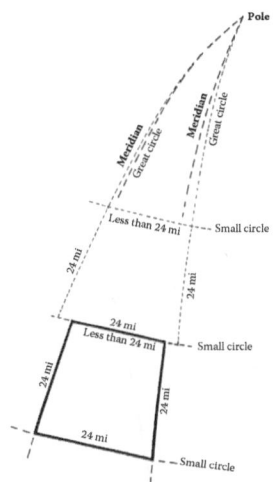

FIGURE 9
Convergence of quadrangles.

FIGURE 10
Convergence of quadrangles.

Colorado, surveys, which were mostly made from about 1860 to about 1880. Since the 1855 *Manual* was in force, the standard parallels were surveyed out 30 miles to the south and not 24 miles as shown above! Ultimately the "square" was then closed with a true meridian (called guide meridians) run north every 24 mile intervals, from one standard parallel to the next. It is now critical to observe 2 items:

- By definition, for each quadrangle the monuments defining the southern and western boundaries are to be accepted and thus control the interior.

- At the standard parallels or correction lines there are now two different monuments, the <u>standard corners for surveys going north</u> and the <u>closing corners for surveys coming in from the south</u>. (See Figure 11.) For the dubious reader, it should be pointed out that both standard and closing corners can be seen from the air on any flight across the midwestern states, especially if the survey lines are accentuated by roads (see Figure 31), the survey pattern in the prairie states. Since the monument(s) control, it is of critical importance for a landowner north of the parallel to have the property tied to a standard corner; land to the south of the parallel, on the other hand, would be tied to a closing corner put Figure 11 here on very near.

Laying Out the Townships

With the boundaries of the quadrangles in place, the next step was to subdivide the interior into 6 × 6 mile townships, with the intent to propagate

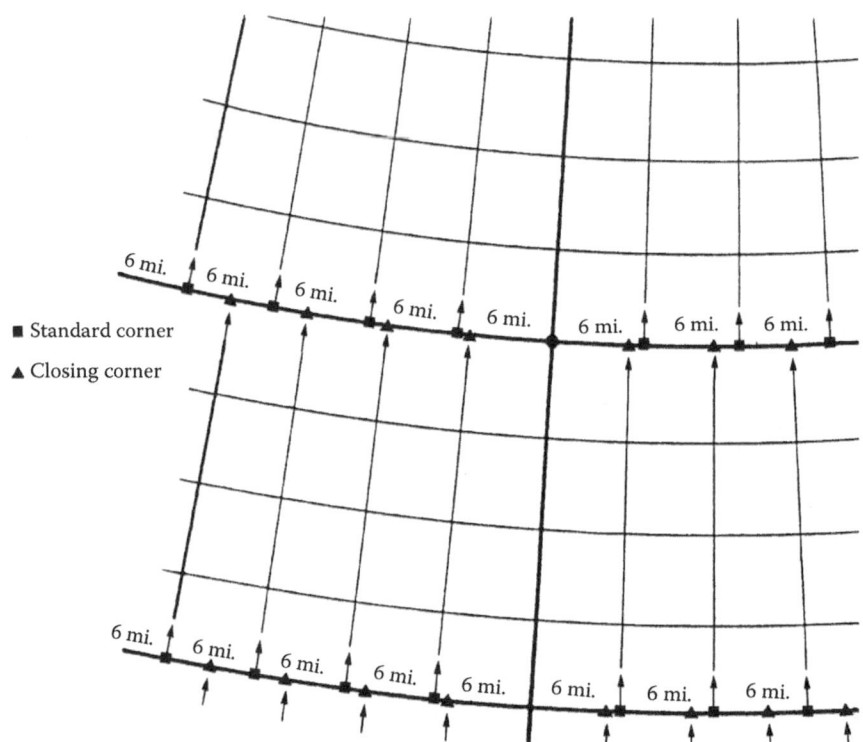

FIGURE 11
Standard and closing corners.

the surveying errors to the north and east as shown in Figure 12. The reader should pay close attention to the surveying (numbering) sequence of the lines shown in Figure 13 below, as well as the fact that most lines were run twice. For example line #2, chained for all its 6 miles, was first run as a random line and then rerun as a finished line #3, so that monuments could be set at 1 mile (80 chain) and/or at ½ mile (40 chain) intervals.

Once again it is now critical to observe that, by definition, the monuments set along the southern and eastern boundary of each township are accepted (i.e., fixed) and thus control the interior surveys.

With this pattern complete, townships can now be uniquely located like in a coordinate system. Starting at the principal point, townships (6 × 6 mile areas) are labeled consecutively by ranges east or west and <u>by townships</u> north or south. Again, the reader is reminded that Figure 14 below is idealized (i.e., it shows no errors, and it illustrates only a limited range of ranges and townships, whereas in reality very large areas were covered by some surveys as shown in Figure 7. As an example, Figure 14 also illustrates the location of three townships: land located in Township 3 South, Range 6 East, of the "X" meridian, or in the abbreviated form as: T3S, R6E, X meridian,

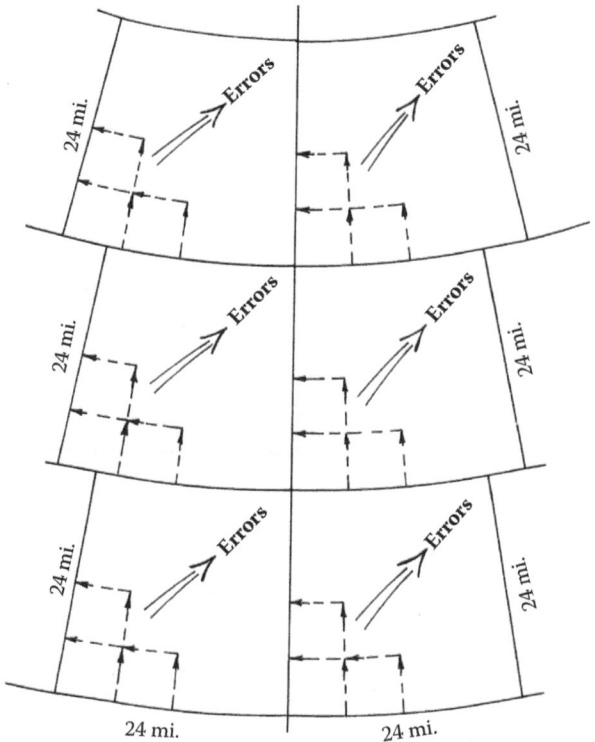

FIGURE 12
Distribution of errors in quadrangles.

containing 36 sq. mi., more or less: T7N, R 4 W, X meridian and T7N, R5E, X meridian, also each containing 36 sq. mi., more or less. As required by the *Manual of Surveying Instructions*, the pattern and labeling was surveyed out to a greater or lesser extent from each principal point. Especially when viewed from the air, this regular "quilt pattern" extends across the plains of North America as far as the eye can see. For example, in the mountains of Colorado and in the more remote areas of Arizona, surveying with transit and chain was difficult at best. In order to fill gaps, half ranges and half townships had to be inserted. To the west of the Holy Cross wilderness area in Colorado, Range 81½ W had to be inserted (see Figure 32).

Laying Out the Sections

Again, without going into details at this time, each township was next subdivided into 1 mile "squares," called sections, by working the surveying errors from the SE to the NW (see Figure 15 and Figure 16). As pointed out previously, it is again very important to carefully study the numbering sequence of the lines in Figure 16, as well as the fact that most lines were run twice. For

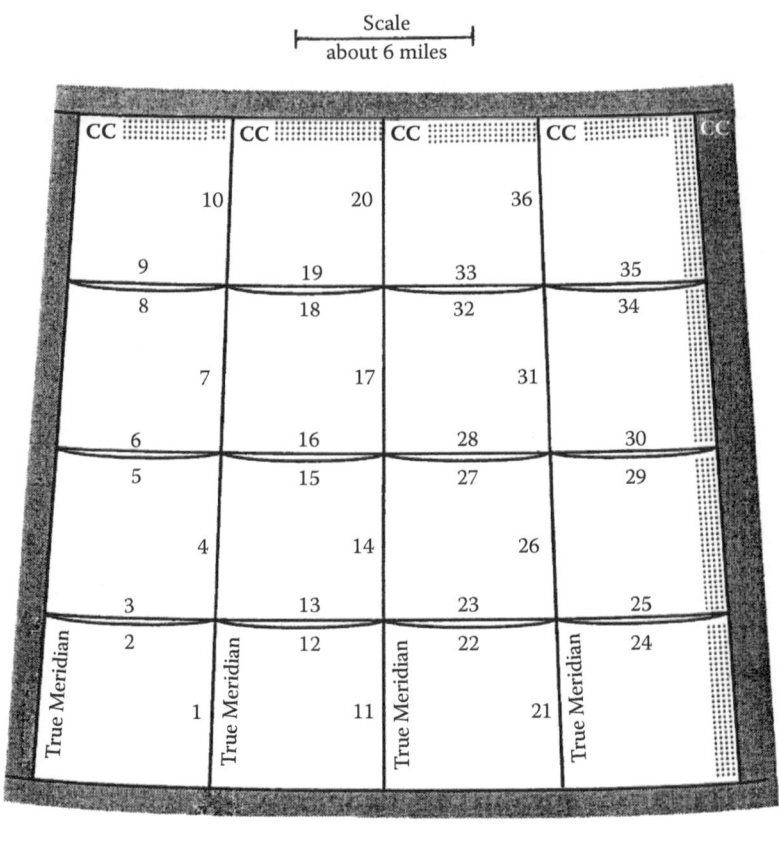

FIGURE 13
The subdivision quadrangle into 6 × 6 mile townships.

example, the 1 mile long line #2 was first chained as a random line and then rerun as a finished line #3, so that the quarter corners could be set on line at 1/2 mile (40 chains) intervals. The original surveyors walked and chained at least 95 miles with a Gunter chain at $2/day, to just survey one township, out of many thousands surveyed!

Following the pattern shown above, the errors accumulated on the north and west side of each township. Thus, the northern tier of U.S. sections (1 to 6) are seldom perfect in a north–south direction, and the western sections (sections 6, 7, 18, 19, 30, and 31) have errors due to convergence on their west side. All these sections can be either shorter or longer than one mile (80 chains), and there are townships where section 6 has completely disappeared. Many illustrations show only a perfectly square township with section numbers. Figure 25 named "Idealized Section Layout" is conventional; however, the

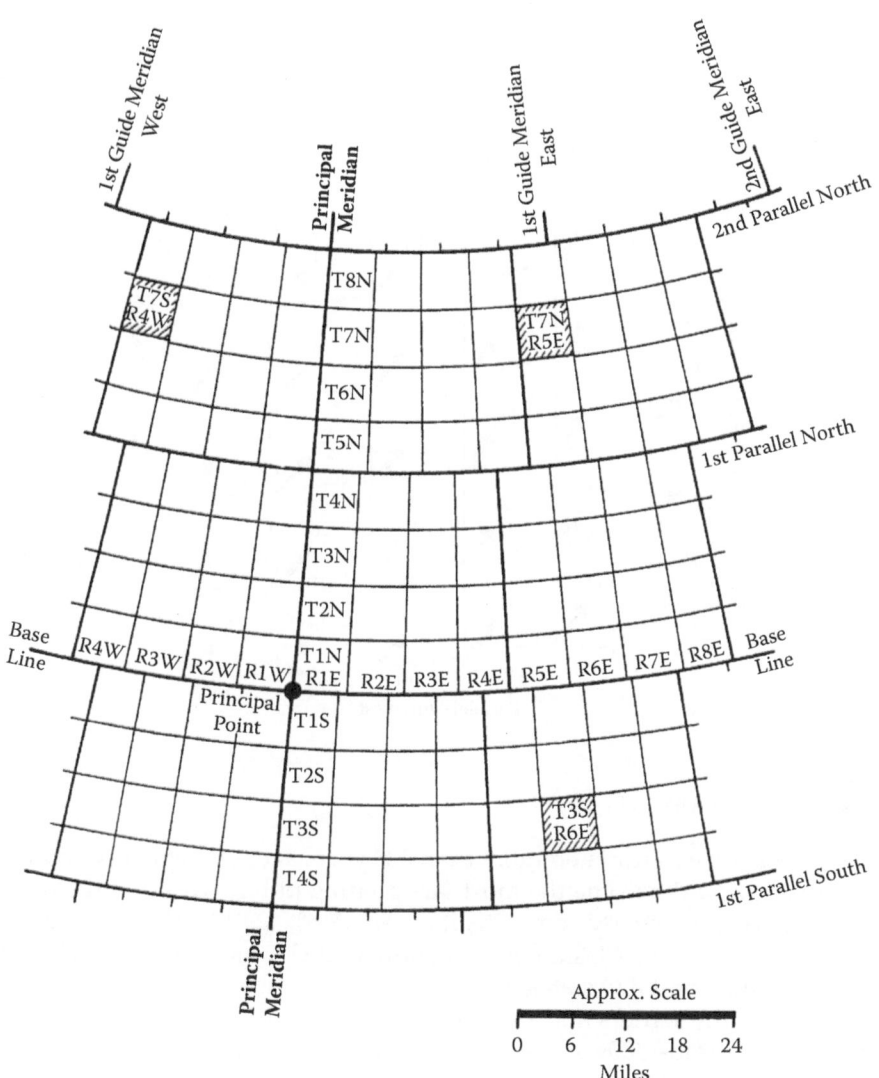

FIGURE 14
Identification of townships by township and range.

"zones of possible closing errors" have been added. Across North America the picture is somewhat different, as shown below, with section numberings for the Standard American (Figure 17) and the Canadian sections (Figure 18) in a boustrophedonic pattern (Greek for "as the ox plows"), Figure 20, and the somewhat unusual Ohio section numberings (Figure 19).

The paragraph following may only be of interest to a surveyor, but it is now time to reflect on the extent of the public surveys. When the Public Land

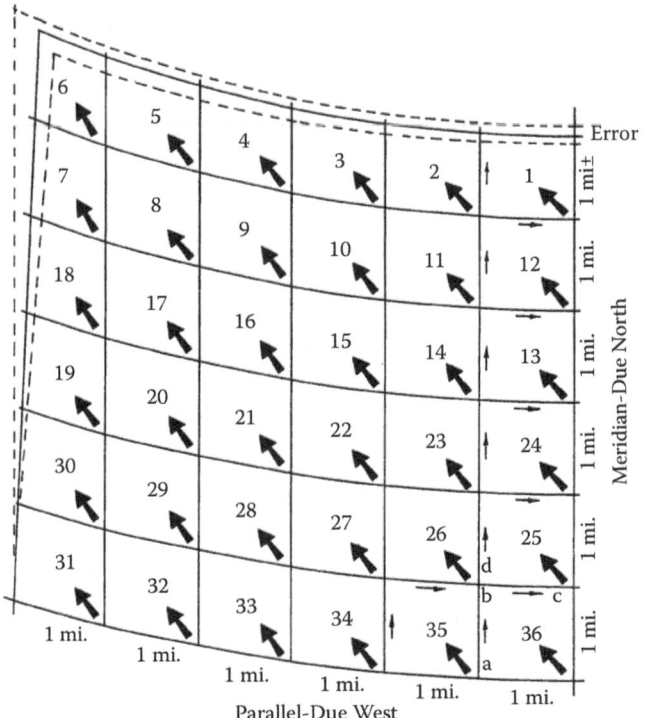

FIGURE 15
Subdivision of a township into sections.

Survey System was initiated more than 220 years ago, probably nobody realized that it would become the most far-reaching public works on the face of the earth. By now nearly complete, it covers 79.5% (2,812,492 sq. mi., including 570,374 sq. mi. in Alaska) of the continental U.S. as compared to 723,786 sq. mi. for the non-GLO states.

Preceding the settlers and railroads, when most of the work was done in the early and mid-1800s, none of the surveyor generals could have realized that more than 78,000 townships, with over 7,400,000 miles (about 300 times around the equator) of survey lines had to be walked and chained by hand. A modern surveyor probably worries mostly about finding a specific section corner, rather than reflecting on whether in all those townships all 1,950,000 interior section corners were set.

Section Corners and 1/4 Corners

The most important end result of the Government surveys is the section corner (or 1/4 corner, if it was set), because in the GLO states the private metes and bounds property surveys have to be tied to it. The principle of naming a section corner (monument) is shown specifically in Figure 21,

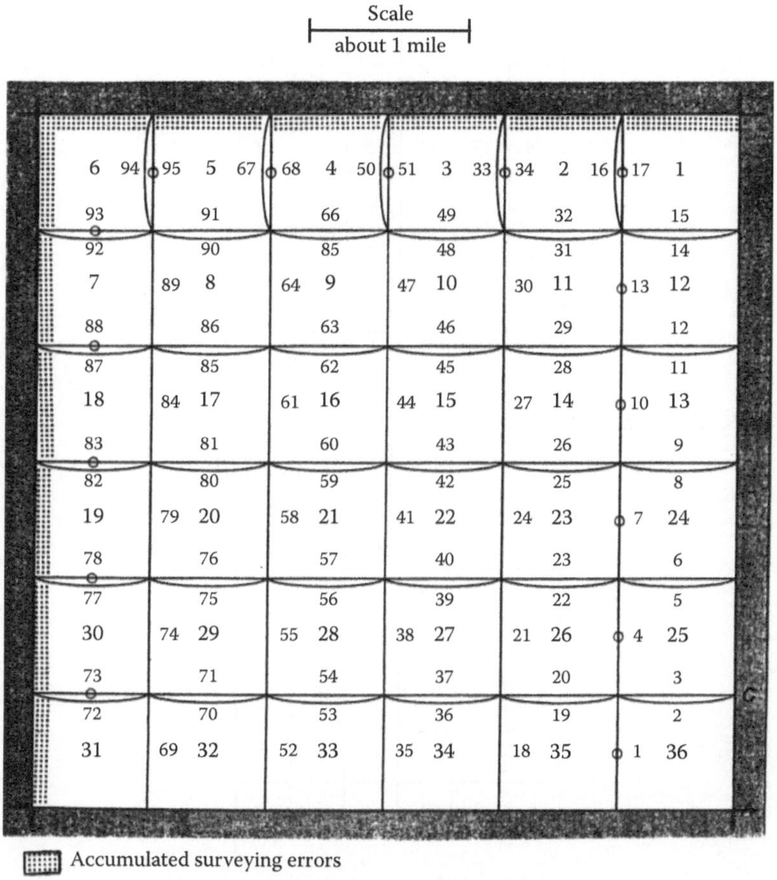

Scale

about 1 mile

6	94	95	5	67	68	4	50	51	3	33	34	2	16	17	1
93		91			66			49			32				15
92		90			85			48			31				14
7	89	8		64	9		47	10		30	11		13	12
88		86			63			46			29				12
87		85			62			45			28				11
18	84	17		61	16		44	15		27	14		10	13
83		81			60			43			26				9
82		80			59			42			25				8
19	79	20		58	21		41	22		24	23		7	24
78		76			57			40			23				6
77		75			56			39			22				5
30	74	29		55	28		38	27		21	26		4	25
73		71			54			37			20				3
72		70			53			36			19				2
31	69	32		52	33		35	34		18	35		1	36

▦ Accumulated surveying errors

Sequence of Surveys for the Subdivision of
a Township into Sections
Note: Total distance chained = 95 miles more or less.

FIGURE 16
The subdivision quadrangle into 6 × 6 mile Townships.

where section 15 is used as an example in an idealized layout. Section corners are named with respect to the section; shown are the NW corner, SW corner, and NE corner of section 15 only. The SE corner has been omitted for clarity.

As shown in summary in Figure 22, a section corner can actually have four names; Figure 21 indirectly shows that in the Unites States, the SW corner of Section 15 is also the SE corner of section 16, the NE corner of section 21, and finally the NW corner of section 23; with all sections located within the same township.

The 1/4 corners derive their name from the fact that lines surveyed (drawn) to opposite 1/4 corners would divide the section into quarters (i.e.,

6	5	4	3	2	1
7	8	9	10	11	12
18	17	16	15	14	13
19	20	21	22	23	24
30	29	28	27	26	25
31	32	33	34	35	36

FIGURE 17
Standard U.S. section numbering.

31	32	33	34	35	36
30	29	28	27	26	25
19	20	21	22	23	24
18	17	16	15	14	13
7	8	9	10	11	12
6	5	4	3	2	1

FIGURE 18
Standard Canadian section numbering.

6	5	4	3	2	1
12	11	10	9	8	7
18	17	16	15	14	13
24	23	22	21	20	19
30	29	28	27	26	25
36	35	34	33	32	31

FIGURE 19
Section numbering in some parts of Ohio.

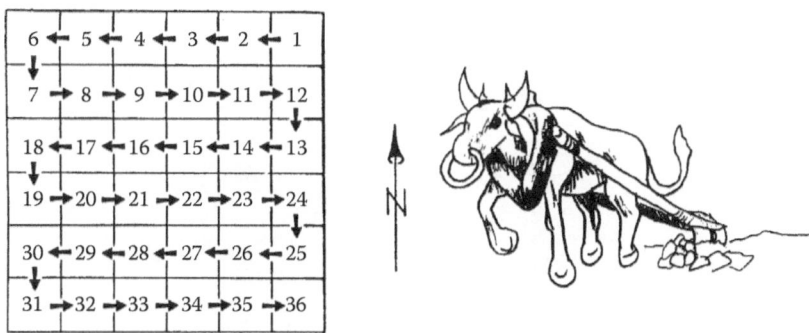

FIGURE 20
Section numbering by bustrophedonic pattern.

connect the N 1/4 corner with the S 1/4 corner and connect the W 1/4 corner with the E 1/4 corner). Figure 21 shows both the W 1/4 corner and N 1/4 corner, the other two 1/4 corners have again been omitted for clarity; 1/4 corners are named with respect to the section, and Figure 23 also shows that 1/4 corners have 2 names. More to follow on the type of monuments set and how to find them.

Once set by a deputized government surveyor and documented by corresponding field notes, a section corner or a 1/4 corner are said to be correct. The monument's position can only be changed by a subsequent independent government resurvey. This is in contrast to monuments set as part of a metes and bounds property survey, where the position of each monument depends on the position of each adjacent monument. As the "quilt pattern" of interlocking surveys expands, the positional errors are distributed randomly throughout the system.

Other GLO Corners Set

In addition to section corners and 1/4 corners, other monuments were set by the Government surveyors, such as: meander corners, closing corners on reservations, grants, and at state boundaries.

> **Meander corners** are set where a survey line intersects a meanderable body of water. The definition of a meanderable body of water has changed over time; a lake 25 acres or larger (1855 *Manual*), a lake 50 acres or larger (1973 *Manual*), above the high water line of a navigable river, nonnavigable rivers wider than three chains, at the bank of a bayou, above the mean water line of an island, etc.

> The Army Corps of Engineers is charged with the designation of a navigable river. It is defined as a river "which is, or has been used as a highway of commerce." The Corps also designates the "point of navigability."

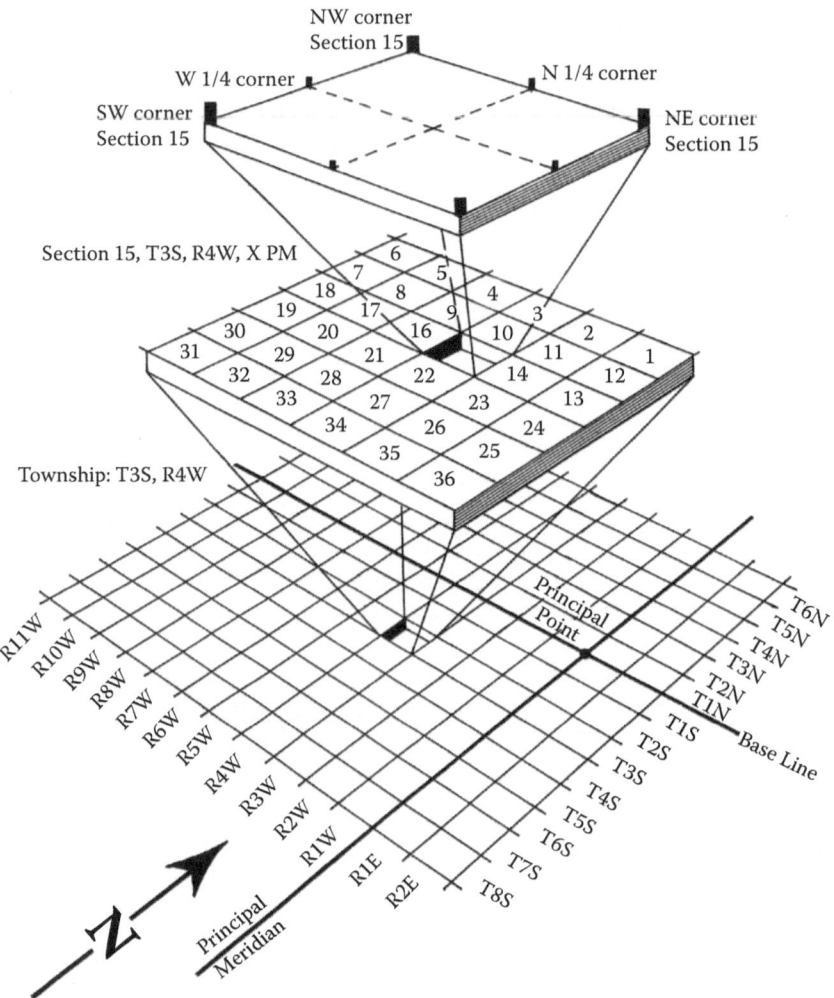

FIGURE 21
Location of section and quarter corners in a section.

Closing corners are set at the intersection of reservations, grants, and state boundaries.

In order to correct fraudulent (defective) GLO surveys, the government has two methods for corrections:

1. Dependent resurvey, and/or
2. Independent resurvey

Note: As with all government surveys, including resurveys, it must be understood that every effort is/was made to respect preexisting

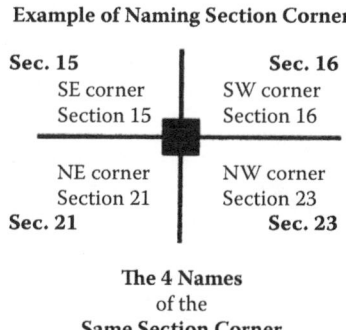

FIGURE 22
The four names of the same section corner.

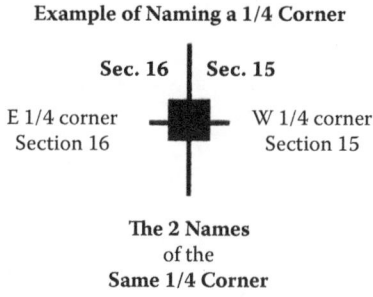

FIGURE 23
The two names of the same quarter corner.

legitimate private property, by including it in the resurvey—see note following.

Dependent resurvey. As the name implies, if possible, all previous monuments are accepted. CAUTION: There is often an appreciable time delay (in years!) between the survey and the signature and final recording (i.e., legitimizing, by the Federal Attorney General). For examples, see: Figure 33 to Figure 36.

Independent resurvey. All previous monuments are voided and the resurvey controls, provided the resurvey has been adjudicated, i.e., signed and recorded by the Federal Attorney General.

The GLO surveys and their interaction with "preexisting private" surveys.

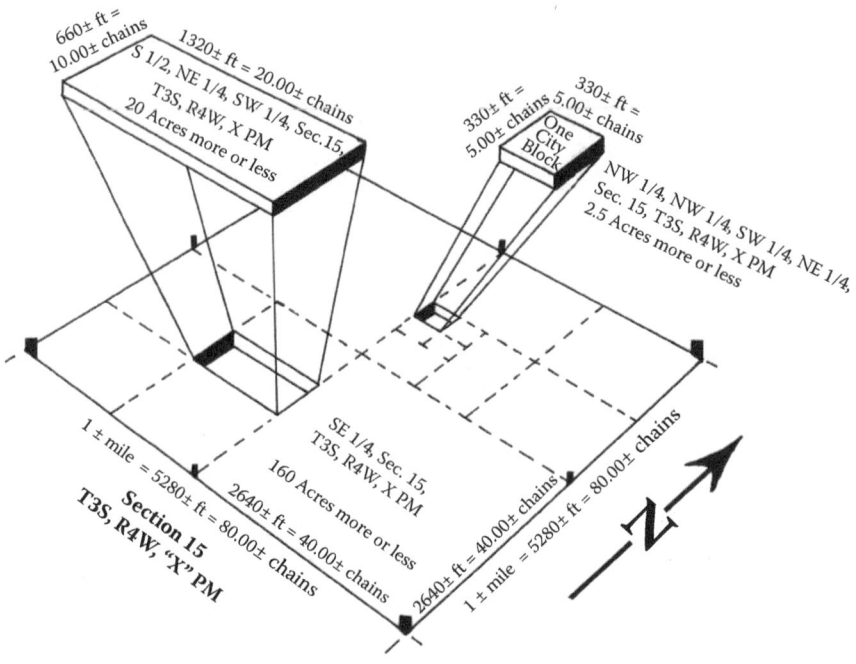

FIGURE 24
Subdivision of a section into aliquote parts.

Throughout their history the Federal Government Surveys interacted with private metes and bounds land claims from French, Spanish, Mexican and Mining claims as well as "squatter rights," and nominally every effort was made to respect legitimate rights. Generally the private landowners were asked to substantiate their land claims through physical monumentation and "legal documentation." Ultimately the land was adjudicated or patented through American judicial court action and the rectangular survey was abutted to preexisting private land. Land parcels, both private and government land, were assigned consecutive lot numbers within a township (6 × 6 miles); full sections (640 Ac) either followed the conventional pattern (i.e., sections from 1 to 36 with lots for the partial parcels of land) or were numbered as lots; thus section numbers above 36 are possible. For an example see Figure 43 to Figure 45, for Township 9 North, Range 10 East, Louisiana Principal Meridian, which has full sections (640 Ac) numbered from 38 to 55. Following the customs of Spanish and Mexican law, the mineral rights, i.e. gold, silver, etc. and often salt, were retained with the surface landownership and thus are not open to mineral prospecting, a point of law especially applicable in the mineral regions of California, Arizona, and New Mexico.

The Aliquote (Area) Subdivision of a Section

Every attempt was made to subdivide a section into regular aliquote parts, where an aliquote part is defined as dividing a number (area) evenly without leaving a remainder.

Irregular sections were first broken into aliquote parts, and the remainder was then subdivided into "government lots" (see Figure 30).

Subdividing the area of a section into aliquote parts is often a source of misunderstanding. There are two parts:

1. Aliquote description.
2. Area associated.
   a. The aliquote description of an area is conceptually straightforward; it *starts with the smallest parcel* and works progressively outward, until it becomes part of the section. The description is usually made in 1/4 of ..., but it can also be made as 1/2 of ... In turn it is necessary to then identify the Township, Range, and Principal point. With reference to Figure 24, three areas are described, first as a full description, to be followed by the "shorthand" version normally seen.
      i. The south-east quarter of Section 15, Township 3 South, Range 4 West, Old Crow Principal Meridian; SE 1/4, Sec.15, T3S, R4W, Old Crow PM, containing 160 acres more or less.
      ii. The south 1/2 of the north-east quarter of the south-west quarter of Section 15, Township 3 South, Range 4 West, Old Crow Principal Meridian; S 1/2, NE 1/4, SW 1/4, Sec.15, T3S, R4W, Old Crow PM, containing 20 acres more or less.
      iii. The north-west quarter of the north-west quarter of the south-west quarter of the south-west quarter of Section 15, Township 3 South, Range 4 West, Old Crow Principal Meridian; NW 1/4, NW 1/4, SW 1/4, NE 1/4, Sec.15, T3S, R4W, Old Crow PM, containing 2.5 acres more or less.
   b. The area associated with the description is more subtle. Even though in most cases every effort was made to have a perfect survey, there are no perfect sections, i.e., exactly 1 × 1 mile.

The GLO surveyors set section corners and depending on the region and instructions in the *Manual of Surveying Instructions*, also set 1/4 corners (N 1/4 corner, etc.). The 1/4 corners derive their name from the fact that lines surveyed (drawn) to opposite 1/4 corners would divide the section into quarters (i.e., connect the N 1/4 corner with the S 1/4 corner and connect the W 1/4 corner with the E 1/4 corner). When established, the original surveying notes show whether or not the 1/4 corners were placed in the middle (40 chains = 2640 ft) between the section corners, or what proportion was used, for example: 40.00 chains from the SW corner and 39.02 chains from the NW corner. For the non-surveyor the consequence of the statement above is that the "quarter of a section" on the ground is <u>not exactly</u> 160 acres, but 160 acres more or less. Since the original monuments and original notes control, it is possible to calculate the area, either based on the original survey or on a re-survey.

Private surveyors subdivide the sections beyond the initial quartering, down to 1/256 parts (i.e., 1/4 of 1/4 of 1/4 of 1/4). As can be seen in Figure 24, this is the boundary of one "standard" city block, 330 ft by 330 ft from center-line to center-line of the streets. There are innumerable city blocks of the size shown across the country; one of the exceptions are the blocks in Golden, Colorado, where the block is 330 ft. × 330 ft., curb to curb, and the streets are extra.

Again, it must be emphasized that for most sections the aliquote areas are fairly true, but as shown in Figure 15 and Figure 16 in this chapter, and Figure 25, in the northern sections (1 through 6) and the western sections (6, 7, 18, 19, 30, and 31), where the surveying errors have accumulated, the aliquote areas can deviate appreciably, both larger and smaller than theoretically. This is especially true for section 6, which theoretically has 640 acres, but can have only a few acres or it even can be non-existent. See Figure 32 to Figure 35.

Since the ***monument of record* controls**, the question inevitably gets around to:

> *What type of monument was set?* and later,
> *How can one find the monument?*

Type of Monuments Set

And the answer is simple: Read the original survey notes! (For an example, see Figure 28.)

Otherwise, the answer is manifold:

- A lay person usually envisions a metal post with a brass or aluminum cap, appropriately labeled (stamped), with a steel insert for a magnetic locator and sticking up about 6 in. above ground. Correct. There are probably many thousands of these "modern" monuments in the ground.
- More subtle, but again the modern trend, are labeled brass caps set a few inches below the ground or below pavement, inside range boxes. A range box is a metal or plastic collar covered with a lid, set on

grade, that is, like a miniature sewer manhole. Occupational hazards include that the range box fills with stagnant water or becomes the home of a rattlesnake, etc.

- Beyond the modern monuments the answers become vague; however, a few thoughts may be helpful.

Surveyors are very resourceful and tend to use anything "on hand" that appears to be permanent, for example:

Native rocks—partially buried and stood on edge, i.e. the rock rests in an unnatural position (see Figure 29)

Quarried rocks—there are many quarried limestone monuments in Kansas and other places (see Figure 29)

Steel driven into the ground—car axles, some with gears still attached

　　Construction reinforcing bar (rebar), any size and length

　　Pipe, any size, usually 1/2 in. or larger

　　Railroad track stood on end

A cross chiseled into bedrock or concrete

"PK" nail driven into road pavement—a questionable monument!

Wooden hubs; a hub is a surveying stake with a square x-section

　　Red or white oak, walnut, redwood, etc. This type of monument when rotted will leave a square soil discoloration as evidence of the monument.

Surveyor's caps: In general, a rebar with a plastic cap, stamped "RLS" followed by a number, represents a private survey by a Registered Land Surveyor whose registration number is shown. The registration numbers are on file at the state's office for Professional Surveyors.

Locating a Monument

There are two essential approaches to locating a monument:

1. Paper and electronic search—CAUTION: not all paper records have been scanned!
2. Physical search

Paper and Electronic Search

1. It is almost essential to have a copy of the original survey notes to see what and often where a monument was set and then what, if any auxiliary ties, etc., were set. (For an example, see Figure 28.) This does not preclude that some old monuments were later replaced.

2. Fortunately, most states now require that professional surveyors file "monument records" of primarily public monuments (for example, see Figure 38 to Figure 40 for Colorado monument records). The documents describe the monument itself, ties to reference monuments, and sometimes the route to get to it. These records are on file at the State Surveyor's Office and sometimes both at the county (surveyor's office or mapping office) or city level. For examples, see Figure 38 to Figure 40. Unfortunately, these records are often not complete, but they can be "pulled" by anybody as a paper or an electronic record.

3. Maps are of great help for locating public monuments. The standard (1:24,000) U.S.G.S. topographic map shows survey lines in red and located corners with a plus (+) sign. In addition it often shows the best approach route. Electronic maps may or may not be suitable. A modern, electronic way now makes it possible by determining the location of the desired monument on the map (latitude and longitude, or by coordinates) and then uses a GPS receiver as a guide to find the location on the ground. *Caution*: Most recreational (handheld) GPS receivers will give a location only to maybe 30 ft.±. Maybe in the future coordinates of all public monuments will become available for an easy GPS location.

4. Aerial photos offer visual and sometimes historical evidence of a monument location. Depending on the location, photos can be purchased at reasonable cost from both the federal and local governmental agencies. Black and white photos are cheaper and usually easier to interpret than color photos.

5. Satellite photography (Google Earth or Microsoft Bing Earth) can be most helpful, even though the resolution (details) may not be as desired.

Physical Search

There are no "rules" for finding a monument, only suggestions. Most monuments (90% or better) are not lost, just obliterated!

The only principle is: "The surveyor has to hunt and search in the field until he has found the best available evidence of a monument. Time is not a consideration" (Brown and Eldridge 10-18).

With a physical description of a monument in mind, a searcher turns into a detective.

Tools Are Helpful

Pacing: If nothing else, pacing is a "tool" which cannot be lost; pacing is most useful for shorter distances or when tie distances are known. It can limit the search area to probably a 10–20 ft radius.

Magnetic compass: Even though society has littered the country with magnetic debris, a magnetic compass together with pacing is often useful, especially in wooded or hilly terrain. Setting the magnetic declination has to be considered.

Shovel, hoe, rake, etc.: Any hand tool to "feel" the top layer of the soil. If a wooden monument has rotted away, or a physical monument has been pulled out of the ground, digging can irreparably destroy valuable evidence and therefore should be used with caution.

Magnetic locator: The tool of choice for a monument with iron content. Unfortunately, a locator cannot distinguish between the monument and any other metal object. Finding a 1/2 in. diameter steel spike, set in the early 1930s, was considered to be lucky; digging almost 6 ft. down in a gravel road with a post hole digger was, however, not unusual.

Hand mirror: A small hand mirror is very useful for reflecting sunlight across old chisel marks on rocks, which have been overgrown with moss or lichens.

Roads, both paved and unpaved: In the GLO states, usually the centerlines of roads follow section lines. Consider SAFETY FIRST! Before digging at the middle of street intersections, permission from the local highway and police authority is very strongly recommended. Looking for a monument does not provide automatic permission to dig in the road or immunity from traffic.

Fence lines: Many fence lines align closely with monuments, especially crossing fence lines. In flat, open rural areas, fence lines can often be visually prolonged for many miles.

"Brushed" survey lines: Survey lines cut free of vegetation for line-of-sight provide an excellent clue for many years after the survey. Often new growth differs markedly in size and color.

Talk to neighbors: If the original landowner does not know the location of a monument, maybe some of the neighbors, neighborhood children, or "old timers" do. Especially in small towns, a visit to the local surveyor may also be informative.

The right time:

- Locating a stone monument in the dead of winter, with 6 ft of snow on the ground and at 12,500 ft. elevation will not work. Wait until July or August.

- When the reservoir is empty (fall and winter), the section corner is readily accessible on foot; when the reservoir is full (late spring and early summer), the corner is under about 10 ft. of water.

Don't join the "No Clue Society." If the monument is there, it can be found by somebody.

The Subdivision of Land—A Summary

The subdivision of land is a very complex subject, with many changes or variations over time and location. In many locations several concepts may have been or are in use and when in doubt, the golden rule of surveying, of <u>following in the footsteps of the original surveys,</u> is to be used. Given below is a *generalized* abstract in order to obtain an overview. Thus, land can be described by:

a. Reference to natural objects.

b. Metes and bounds.

c. Aliquot parts of the rectangular survey system.

d. Parts of an urban subdivision.

e. Condominium, or three dimensional space.

f. State Plane Coordinates (NAD'27 or NAD'83).

   i. **Description to natural objects.** As the name implies, the description refers to natural objects, such as trees, the shore of a lake, the thread (middle) of a stream or river, etc. It may contain some numerical data, such as the length of a pipe smoke, a half-hour walk, or a day's ride, etc. No modern surveyor would ever write such a description, yet many are still valid and in use today.

   ii. **Metes and bounds.** Probably the oldest form of property description, dating back to Mesopotamia at least to 4,000 BC. A metes and bounds description starts at a point of beginning and describes by direction and length each boundary leg in sequence, as one would walk the property. A very valuable attribute is a description of the monuments and the area enclosed. For more details and an example of metes and bounds description, see below.

   iii. **Aliquot parts of the rectangular land system.** This type of description is only applicable to the Western United States and Canada. Thus, a parcel of land may be described as: NE 1/4, SW 1/4, Sec. 7, T 4 S, R 21 E, Coyote Meridian, containing 40 acres more or less.

   iv. **Parts of an urban subdivision.** Land may be described by lot, block, name of subdivision and filing, for example: Lot 9, Block 3, Last Chance Subdivision, 2nd Filing, Irontown, Green County, Pennsylfornia.

v. **Condominium or three-dimensional space.** The concept of a condominium description actually describes the specific individual three-dimensional airspace together with the elements shared in common with all other condominium owners. For example: An individual may own Rooms 508, 509, and 510, in Building "C," and shares in common the land, bearing walls, roof, stairwells, etc. Normally, by direct reference, condominium owners also share the maintenance costs of the common (i.e., the building, grounds, clubhouse, etc.)

vi. **State Plane Coordinates** (NAD'27, NAD'83 or NSRS'06). For the professional (engineer, surveyor, etc.) property is best described by state plane coordinates. It not only provides the only viable computer base, but also the property points are uniquely and "permanently" fixed in space, and the mathematics is all inclusive (for length, bearings, area, etc.). At least 35 states have passed permissive, but not mandatory legislation for the use of the state plane coordinates system. **CAUTION: Do NOT intermix NAD'27, NAD'83 or NSRS'06 coordinates or convert mathematically from one to the other.**

Supplemental Comments for Figure 25 to Figure 40

In an effort to make the illustrations as large as possible, the captions had to be shortened or in some cases completely deleted. Below are given additional pertinent comments.

Figure 25. Idealized section layout: This illustration started as a standard reference sheet for an office. For increased versatility, section numbers around the core section, the zones of possible closing errors, as well as the square measures, were added.

Figure 26. Colorado's Principal Meridians: Land in Colorado is tied to one of three Principal Meridians; the major and NE part to the 6th Principal Meridian (PM), the SW part by the New Mexico Principal Meridian and only a very small portion to the Ute Principal Meridian around Grand Junction. The "baseline" of the 6th PM passes through the southern part of Boulder.

Figure 27. Townships and ranges of the Denver, Colorado, area. The land around Denver is clearly controlled by the GLO survey system. The "baseline" of the 6th PM at 40° N latitude, surveyed in 1859, passes through the southern part of Boulder with the 1st parallel, 30 miles and 2nd parallel South, 60 miles south, etc. The surveys were governed by the 1855 *Manual of Instructions*. Later manuals reduced the distance of the parallels to only 24 mi each.

Figure 28. Copy of original field notes. Probably the most valuable piece of data for a resurvey. Shows distances (in Gunter chains) between ¼ corners and section corners, as well as the type of original monument set; for example, at 80.00 chains set stone 4 in. × 6 in. × 20 in. Even though the distances are given to the nearest

1/00 chains = 0.66 ft., a modern resurvey will generally not consistently confirm the fraction of the foot. Understandably, in flat terrain the accuracies of these original measurements are better than in the mountains. The variation (declination) measured with a magnetic compass to the nearest ¼ degree, shown at 14° 30′F., was most probably not physically determined on that day (December 8, 1863), but carried forward on paper for probably several days from a night observation. Local magnetic anomalies were seldom observed. The bearings given are all magnetic bearings. The height of the mountain at 50.00 chains is estimated at 500 ft.

Figure 29. Illustration of monumentation (1855 *Manual*). This illustration is included to show the care and effort the GLO surveyors took to make and preserve monuments. Especially during the initial surveys, supplies were very scarce, and the survey crews had to live off the land from food to "naturally" survey monuments. Where available, wood for posts worked well, even though the monuments would rot; stones stood on end made excellent and durable monuments. In some areas, limestone posts were set instead of wood posts. Pits filled with charcoal and mounds were only made when there was no other means to make a "permanent" monument.

Figure 30. Subdivision of an "irregular" section. Especially along the northern and western perimeter of a township it was not always possible to survey out full sections of 640 acres each. The subdivision of partial sections started in the SE corner with a 160-acre parcel, if possible, to be followed by 40 acres (1320 ft. × 1320 ft.) and then 10 acre parcels (660 ft. × 660 ft.) to the north and west. Parcels less than 10 acres were designated as government lots.

Figure 31. Regular section pattern in Nebraska. An excellent black and white picture (map) of the regular "square" checker pattern of the GLO survey. The pattern becomes even more obvious when viewed from satellite color photography with the computer; another impressive sight is from a commercial airplane flying across the mid-continent where the pattern appears to stretch from horizon to horizon. Implied on the map is that roads border most townships. When these roads were developed during the horse-and-buggy days, the "dog-legs" on the north–south roads across the standard parallel were preserved, with the result that these intersections now pose a major automobile hazard (i.e., dead man's corner). Not shown on this map, but very visible from the air or satellite, are center pivot irrigation sprinklers, the larger marking each section with a 1-mile diameter circle.

Figure 32. Section layout in mountainous terrain. One of several examples where the ideal paper model fails in real life. Surveying above the timberline and up to 14,000 ft. or higher elevations in Colorado was a challenge for even the hardiest surveyors, and mistakes were made. In this case an E–W void had to be filled in with R 82½ W; townships containing less than 36 sections, and many sections are not "square." Reading the original field notes, consulting USGS topographic and U.S. Forest Service maps, and checking satellite photography, as well as talking to local surveyors, is highly recommended before undertaking any ground surveying jobs. Even with helicopters and GPS, this terrain is accessible in summer for only 3 or 4 months.

<u>Figure 33 to Figure 35.</u> Original and resurvey plats of township surveys. A sequence of three figures showing the progression from an original township survey (made in 1870) to a dependent resurvey (made 1926), and finally the modern (1994) shape. As shown on the modern map (Figure 35), the area is roughly between 5,700 ft. and 8,500 ft. elevation in the foothills of the Rocky Mountains.

The original survey, recorded in 1870, shows a perfect township with perfectly "square" sections. The northern and western perimeter are lotted, indicating that the survey is (was) not perfect. Other than that, the terrain is very rugged; even by modern standards, there is no explanation why Sections 25, 26, 35, and 36 were not surveyed.

Probably responding to complaints from local landowners, the U.S. government conducted a physical dependent resurvey in 1926–27. Whenever possible, existing monuments and private ownership are accepted. Thus section 1 "grew" from about 640 acres to about 1,000 acres, section 5 on the other hand "shrunk" from about 640 acres to about 300 acres; since land can be sold as "so many acres, more or less," this is perfectly legitimate. The southern portion of the township was lotted out into tracts, probably in order to respect private ownership. It should be noted that the field work, with undoubtedly new and shiny brass monuments, remained in limbo and of no value until the plat was finally signed, i.e., adjudicated, about 6 years later in 1930. Surveyors and lay people beware; a new and shiny government surveying brass cap, "officially" stamped, is invalid until the paper document is filed and recorded, i.e., legalized by an appropriate signature(s).

Finally, a careful comparison between Figure 34 and Figure 35 shows that the GLO, 1930 plat, and the 1994 USGS maps in the northern half disagree in many areas. Most noticeable is that a single corner now becomes a double corner between sections 4 and 5, sections 8 and 9, and sections 16 and 17. Money permitting, there most likely will be some additional resurveys.

<u>Figure 36.</u> Dependent resurvey plat. A dependent resurvey in the foothills of the Sierras in California's "gold country." The original plats were approved January 29, 1876, and May 5, 1888. This resurvey is unusual in that photogrammetric processes were used, especially on the north and west boundaries, to "restore corners to their true original location."

<u>Figure 37.</u> Supplemental plat showing mining claims. Initially, a record search for a property in a GLO area starts with the original field notes and the record plat, to be followed by supplemental plats or even notations made on the back of written documents. This supplemental plat shows patented mining claims by claim name and claim number, as well as placer claims by name and number. For a possible resurvey and evaluation of possible claim conflicts, it will also be necessary to obtain the written records of all applicable and neighboring mining claims. A novice should be cautioned that the location of claims shown on paper may not agree with the location of the same claim on the ground.

<u>Figure 38 to Figure 40.</u> Colorado Land Survey Monument Record—Filing Instructions, Master Index and Monument Record. The three pages have

been included as an example of a very effective public monument record system. Originally only on paper but now electronically, a surveyor can "pull the record" to help locate a specific GLO monument (i.e., section and ¼ section corner). A future improvement would be the addition of GPS coordinates of the monuments to the record.

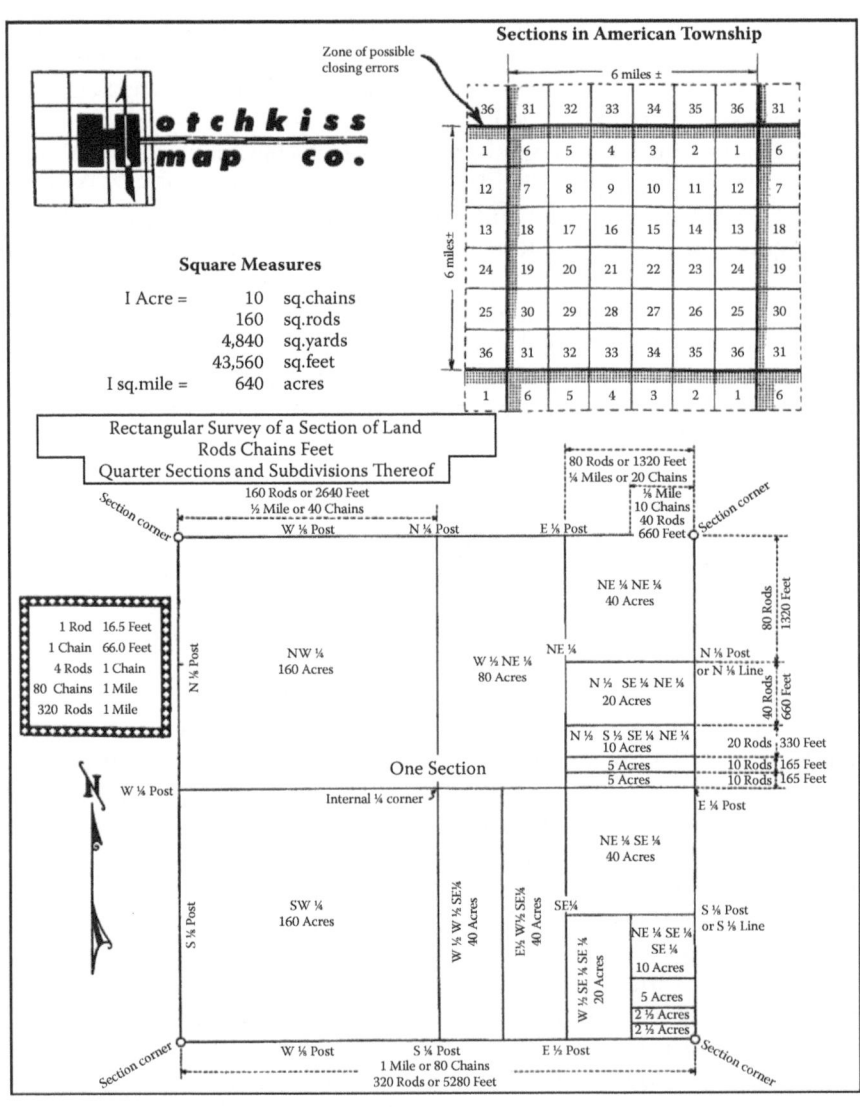

FIGURE 25
Idealized section layout.

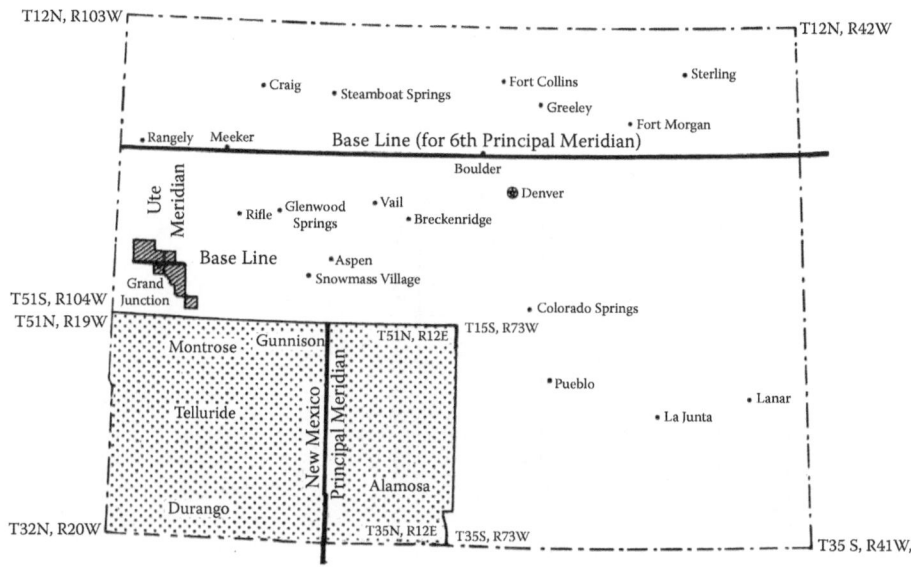

FIGURE 26

Colorado's Principal Meridians (6th Principal, New Mexico, Ute).

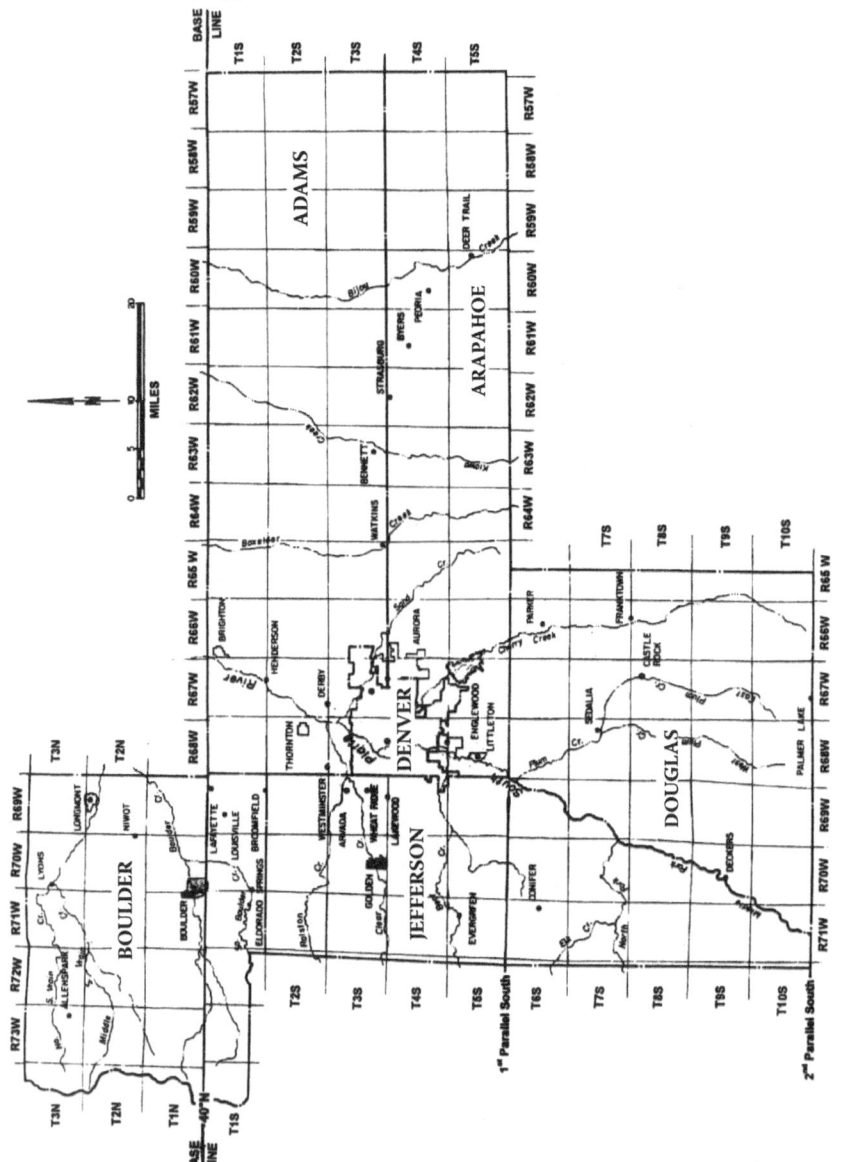

FIGURE 27
Extent of Township and Ranges, Denver, Colorado Area.

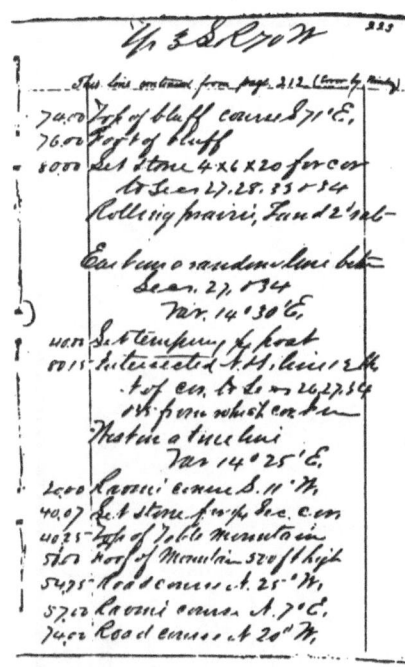

Tp3 SR70W

This line continued from page 212 (Error by ----)

74.00	Top of bluff course S 71°E
76.00	Foot of bluff
80.00	Set stone 4x6x20 for cor
	to Secs 27, 28.33 & 34
	Rolling prairie, Land 2nd rate
	East and on random line between
	Secs. 27 & 34

Var 14°30'E

40.00	Set Temporary ¼ post
80.15	Intersected N–S line 12 lks
	N of cor. to Secs 26, 27, 34
	& 35 from which cor. I run
	West on a true line.

Var 14°30'F

20.00	Ravine course S 11°W
40.07	Set stone for ¼ Sec. cor.
40.25	Top of table mountain
50.00	Foot of mountain 500 ft high
54.75	Road course N 25°W
57.00	Ravine course N 7°E.
74.00	Road course N 20°W

Editorial notes:

This page of field notes was selected to show the variety of information, or lack thereof; the number in the left column is in Gunter chain, var = magnetic variation (declination).

(a) At 80.00 chains described stone with dimensions in inches.
(b) At 40.07 chains set stone, size ?
(c) At 80.00 chains started a random line to the East for 80.15 chains; the temporary ¼ corner was then reset on the return trip at 40.07 chains (80.15/2).
(d) At 80.00 chains note reference to land quality.
(e) At 50.00 chains the reference is to a 500 ft high mountain, the modern value is 541 ft.

FIGURE 28
Example of original filed notes by Pierce, Dec. 8, 1863.

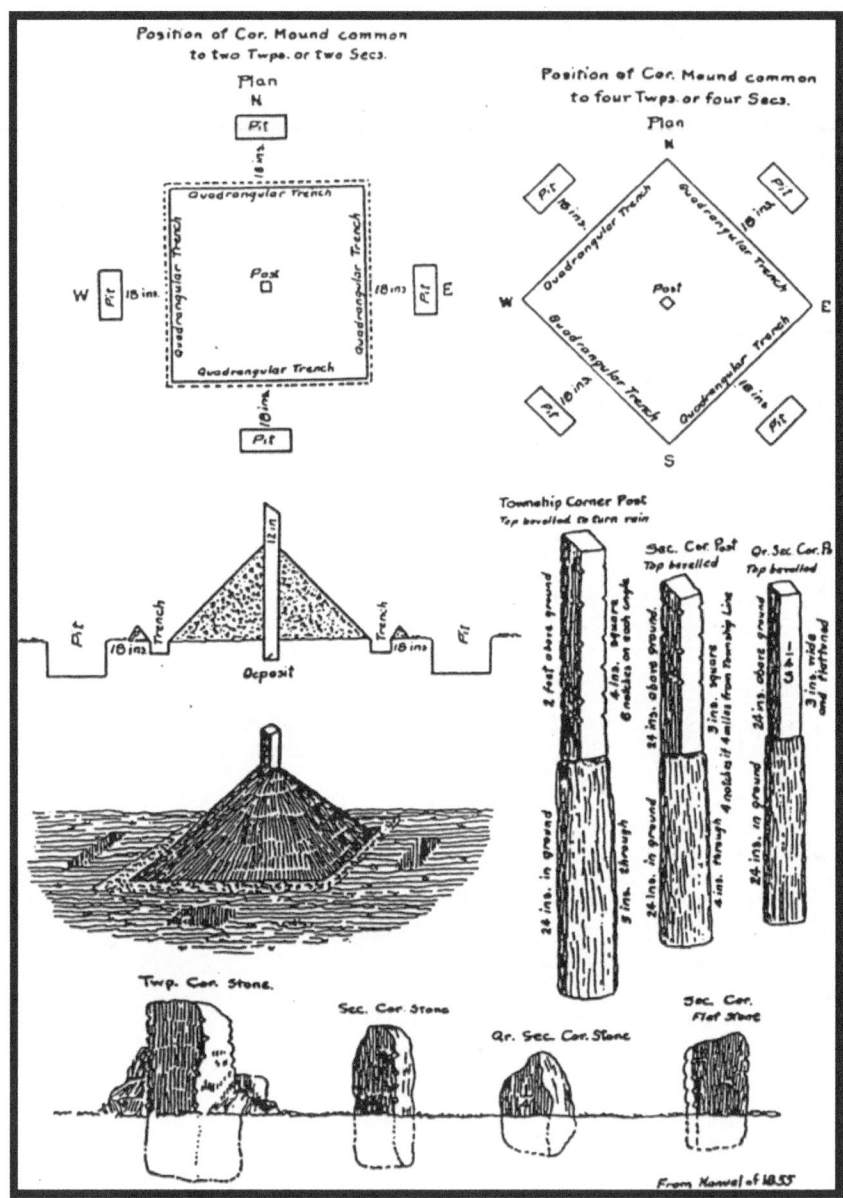

FIGURE 29
Illustration of corner boundaries.

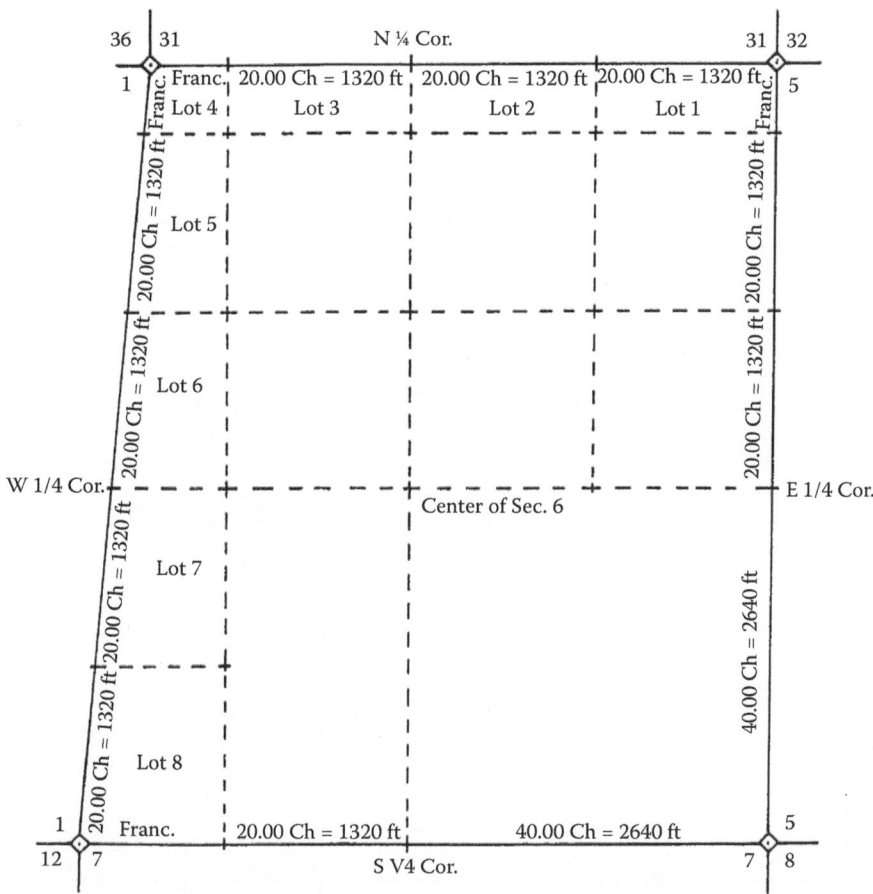

Note:
- The E 1/4 Corner and S 1/4 Corner were set at 40.00 chains from the SE Cor. Sec. 6, resulting in the SE 1/4 corner, the Sec. 6 containing 160 Acres more or less.
- From the E 1/4 corner, the survey continued north for 20.00 chains, and again for 20.00 chains, with the closing error allocated into government lots 1, 2, and 3.
- From the S 1/4 corner, the survey continued west for 20.00 chains, with the error allocated into government lots 5, 6, 7, and 8.
- The "center of section" is located at the intersection of cords drawn from opposite 1/4 corners, i.e. from the N 1/4 corner to the S 1/4 corner and from the W 1/4 corner to the E 1/4 corner.

FIGURE 30
Idealized subdivision of an "irregular" section.

Gage County, Nebraska

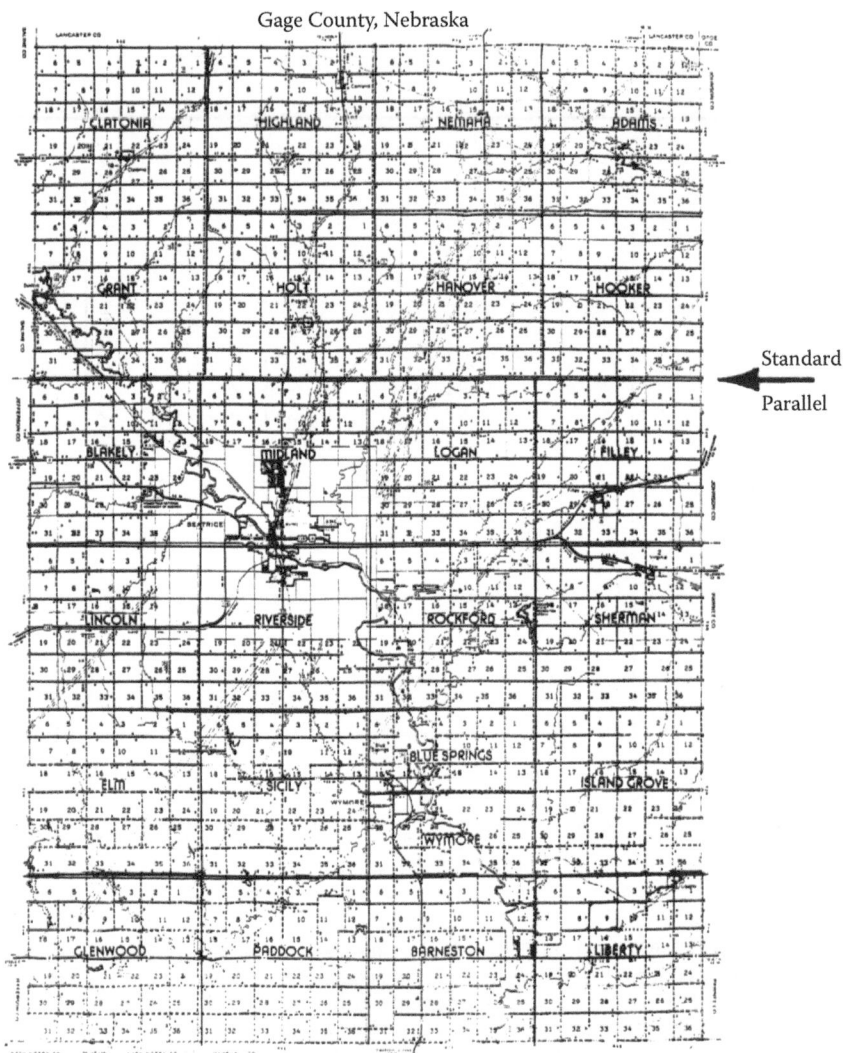

Standard

Parallel

This illustrates the regular section pattern in Gage County, Nebraska.
Somewhat unusual is the naming of each township.
Note: Standard Parallel, with the offset road pattern.

FIGURE 31
"Regular" section pattern in the prairie states.

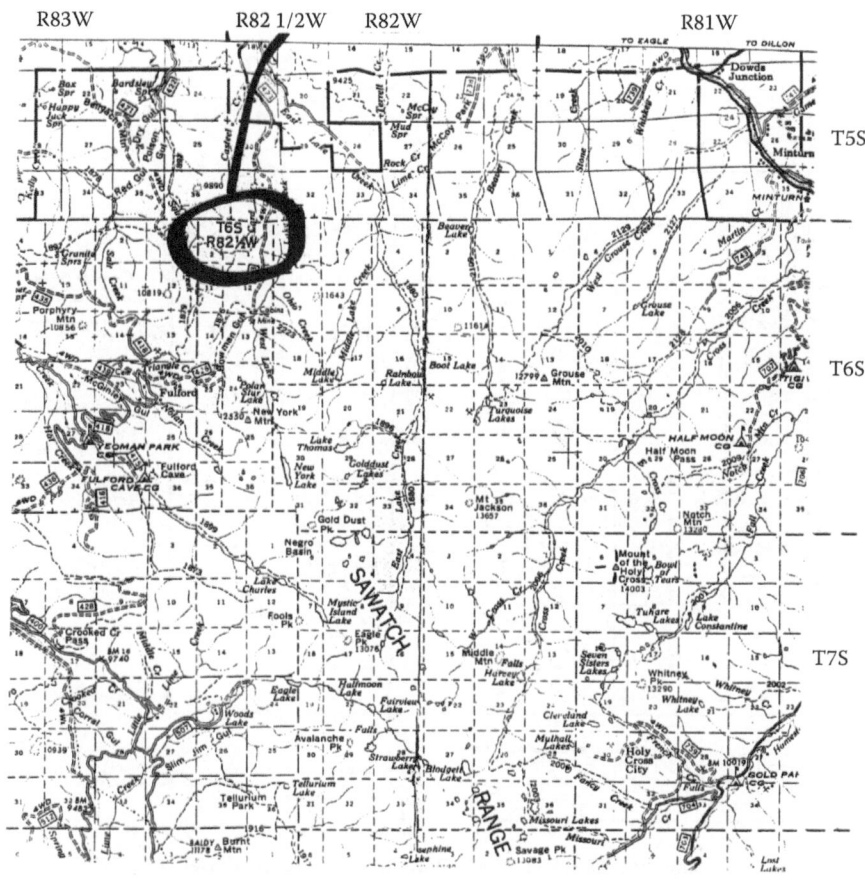

Note:
  – R82 1/2 W has been added to fill a surveying gap.
  – In "T7S", sections 1,2, and 3 are "reasonably" regular;
    section 4 is enlarged and sections 5 and 6 are missing.

FIGURE 32
"Irregular" section layout in mountainous terrain.

Township N°1 South: Range N°71 West of the 6th Principal Meridian

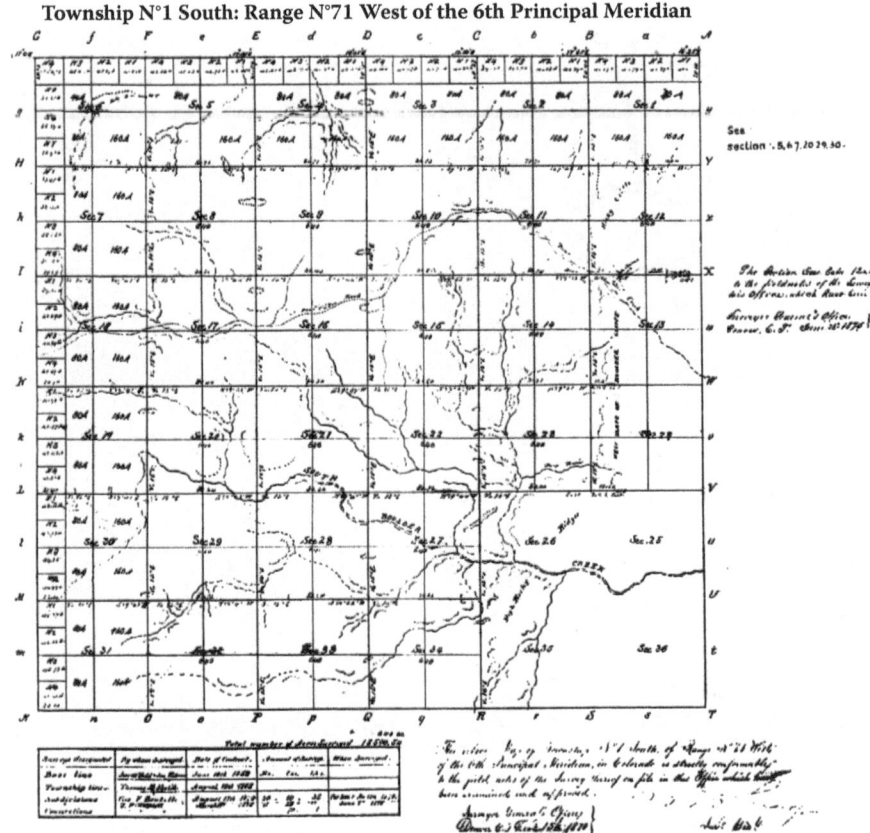

Original Township Plat, T1S, R71 W, 6th PM, filed and Recorded, Dec. 15, 1870.

Note:
  – All but sections 25, 26, 35 and 36 are platted and subdivided.
  – The northern and western boundaries are subdivided into government lots.
  – In light of the subsequent resurvey, the reader should note
    the size of sections 1, 5, and 36.

FIGURE 33
Original township plat.

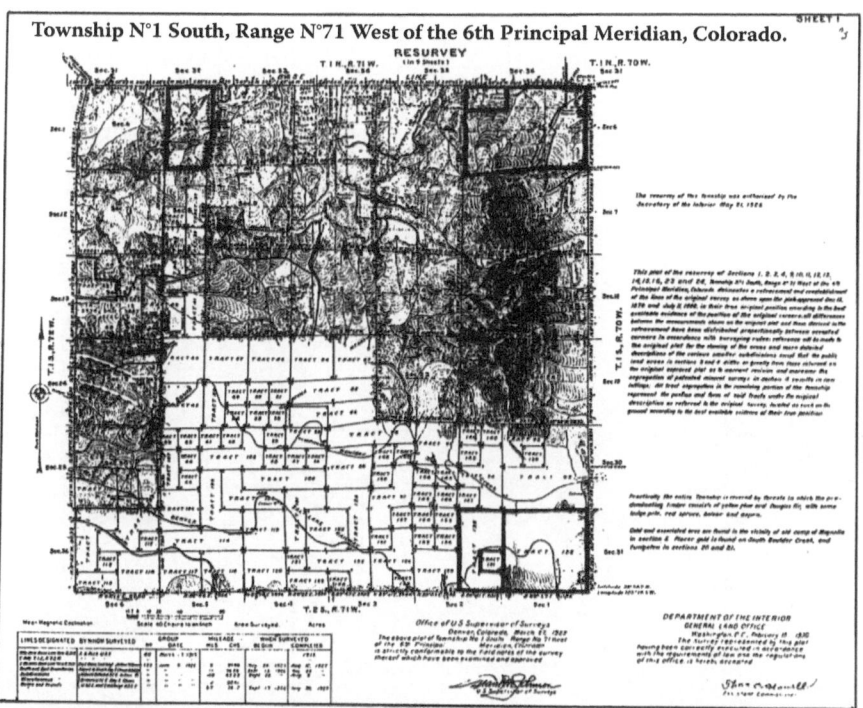

Resurvey Plat of Township, T1S, R71W, 6th PM, Filed and Recorded, February 19, 1930.

Field work mostly 1926–27, signed by Denver Office March 27, 1929,
signed by General Land Office, Washington DC, February 19, 1930

Note – compare to the original plat.
– Section 1 and Lot 4 in section 1 have been considerably enlarged.
– Section 5 has been reduced to about ½ its original size.
– The southern half of the township, including section 36, has been subdivided into
government lots.

FIGURE 34
Resurvey plat of township.

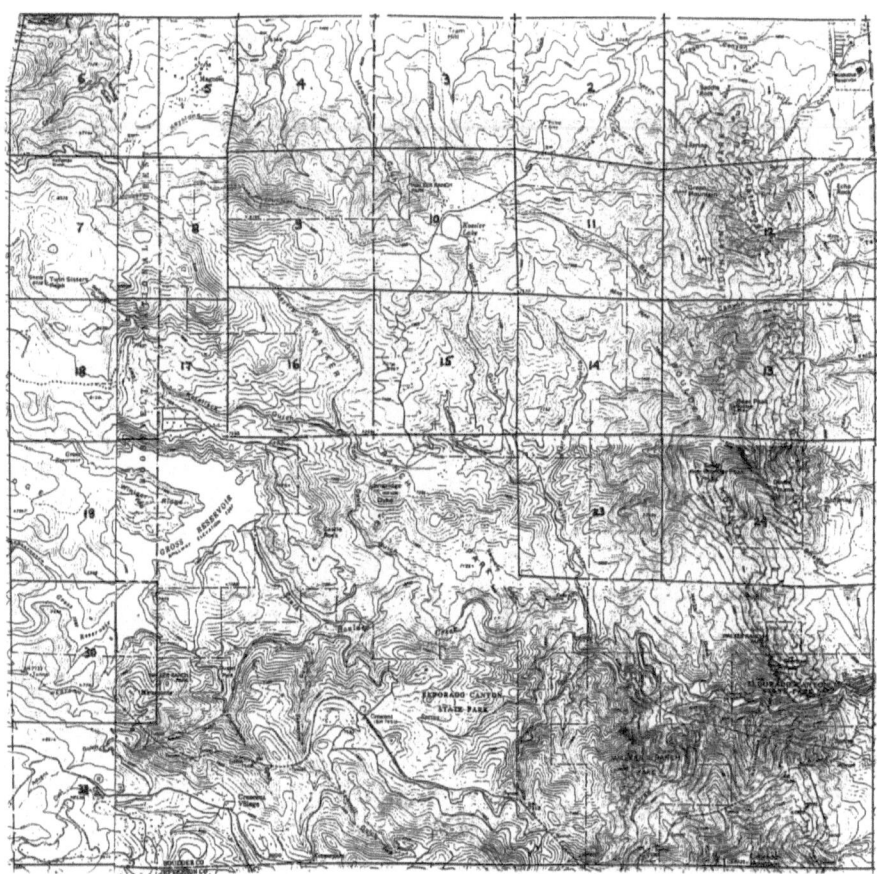

Compared to the original plat and the resurvey plat the actual surveys, as shown by the U.S.G.S. maps (Eldorado Springs and Tungsten 1994), differ appreciably.

Editorial note: the section lines have been enhanced by hand.

FIGURE 35
The "modern" (1994) shape of township, TIS, RITW, 6th PM.

Township 18 North, Range 8 East, of the Mount Diablo Meridian, California.
Dependent Resurvey

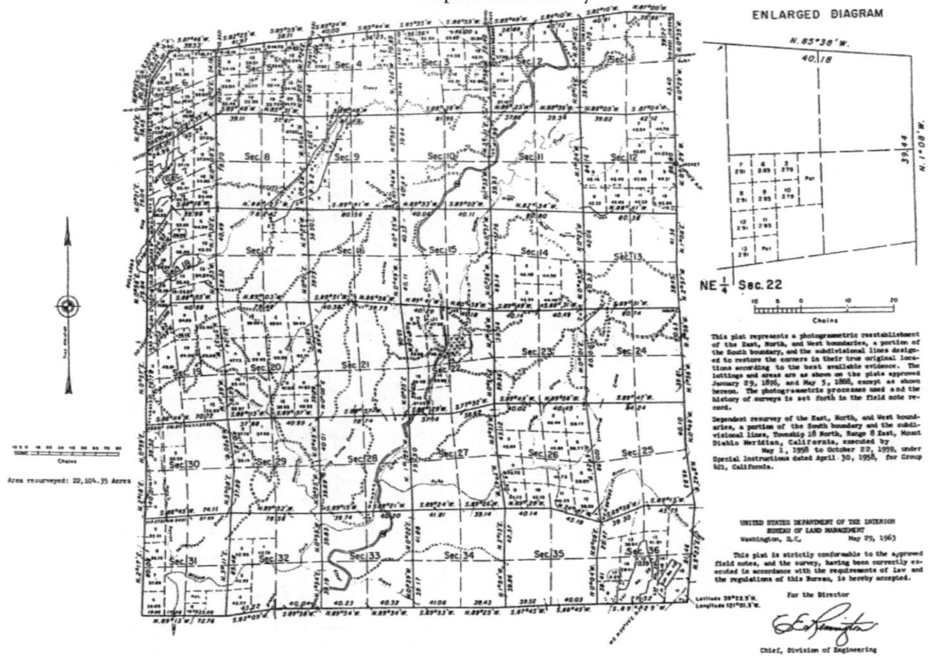

Original surveys made in the 1870s; resurveys May 1, 1958 to October 22, 1959; plat approved May 29, 1963.
Vegetation: heavy stands of yellow and sugar pine, white and douglas fir and cedar.
Topography: foothills of the Sierras, elevation 2500–3000 ft.
Area of 6×6 mi. "square" township = 23,040 acres; area resurveyed 22,104 acres

FIGURE 36
Dependent resurvey—central California.

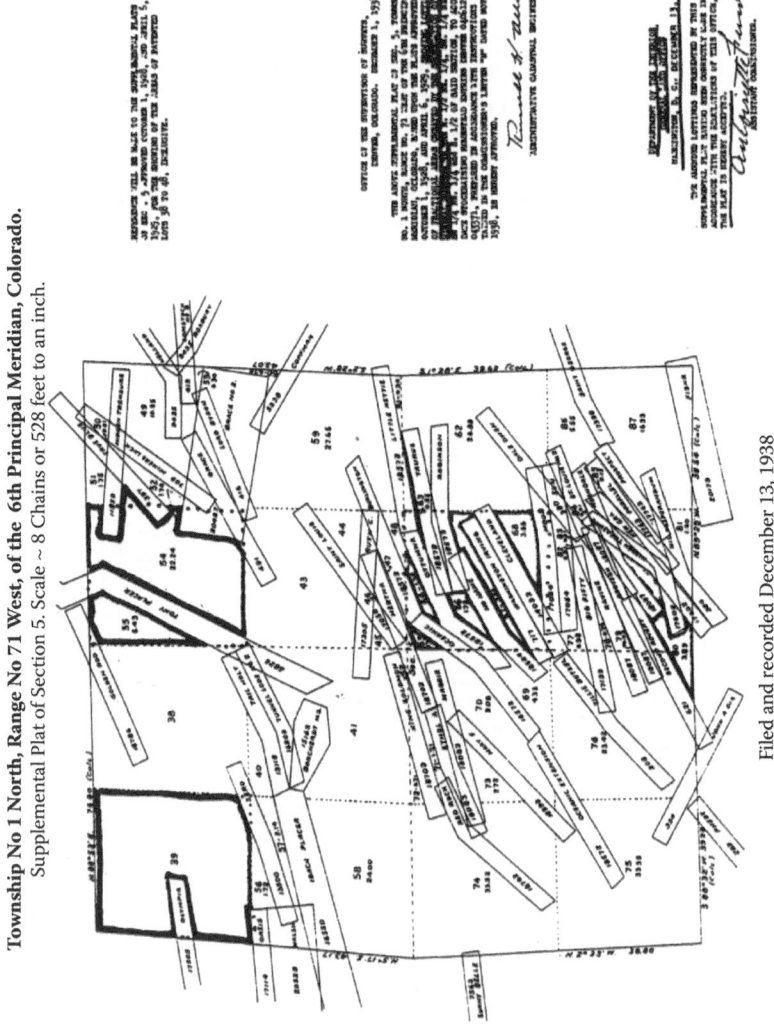

Township No 1 North, Range No 71 West, of the 6th Principal Meridian, Colorado. Supplemental Plat of Section 5. Scale ~ 8 Chains or 528 feet to an inch.

Filed and recorded December 13, 1938

FIGURE 37
Supplemental plat of Sec. 5, T1N, R71W, 6th PM. Showing mining claims.

LEGISLATIVE DECLARATION: It is hereby declared to be a public policy of this state to encourage the establishment and preservation of accurate land boundaries, including durable monuments and complete public records, and to minimize the occurrence of land boundary disputes and discrepancies. Monument Records must be filed pursuant to Title 38, Article 53, Colorado Revised Statutes (1994).

DETAILED INSTRUCTIONS BY ITEM NUMBER ON THE FORM:

Use permanent black lettering and lines which can be reproduced.

1. Indicate type of monument.

2. Describe monument found and accepted (in detail — include size, shape, material, color, and other pertinent markings).

3. Describe monument set by you (in detail — include size, shape, material, color, and other pertinent markings).

4. Make a neat sketch showing the relative positions of the monument and accessories (reference points). Accessories should not exceed 330 feet (5 chains) from monument. Refer to Chapter IV, Section 83 through 114 inclusive, Manual of Surveying Instructions – 1973. Give dimensions which should be *ACCURATE AND NOTED TO AT LEAST 0.1 FEET*; include north arrow; state basis of bearings, if used. Show and describe in detail contradicting monuments. Show street names or highway numbers, if applicable. Indicate scale or state N.T.S. (not to scale). Show markings exactly as found. (A statement "as per Manual" is not sufficient.)

Fill in the "Date of Field Work to Establish or Rehabilitate Monument" (date block "a") and/or the "Date Monument was used as Control (date block "b")." These may be two different dates; for example, the date of the field work and the date the plat was signed. If a monument is established, restored or rehabilitated, date block "a" must be completed. If a monument was used as control, but was not established, restored or rehabilitated, date block "b" must be filled in. The six month filing deadline is calculated using the earlier of the two dates.

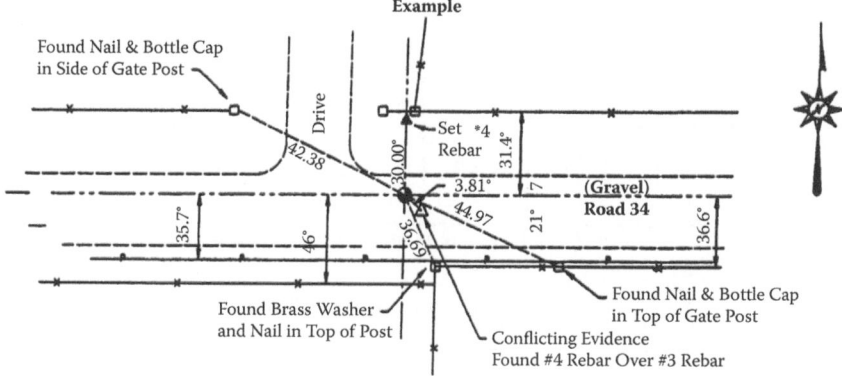

Example

5. Original signature and seal must be on each monument record submitted to the Board for filing. The signature must be through the seal (Board Rule VIII (2) B). Also, list your firm's name, address, and phone number.

6. Show location of monument on section diagram. EXCEPTION: if monument is in an area not covered by the public land survey system, show point loation as near as practical on projected section, township, range, and Principal Meridian. Write "projected" above diagram.

7. Give section, township, range, Principal Meridian, county, and index reference number. Exercise extreme care to ensure that section, township, range, Principal Meridian, county, and index reference number all correlate. Index reference number must be assigned by the surveyor. See sample index sheet for instructions. EXCEPTION: if monument is in an area not covered by the public land survey system see item #6 above.

8. If the monument is on a county boundary, reference, all adjoining sections, townships, ranges, and Principal Meridians within appropriate counties (if monument is on a county line which is also a range line or township line, the index reference number will be different in the various counties). This item is to be used only if the monument is located on the county line.

FIGURE 38
Example of Colorado Monument Record, page 1.

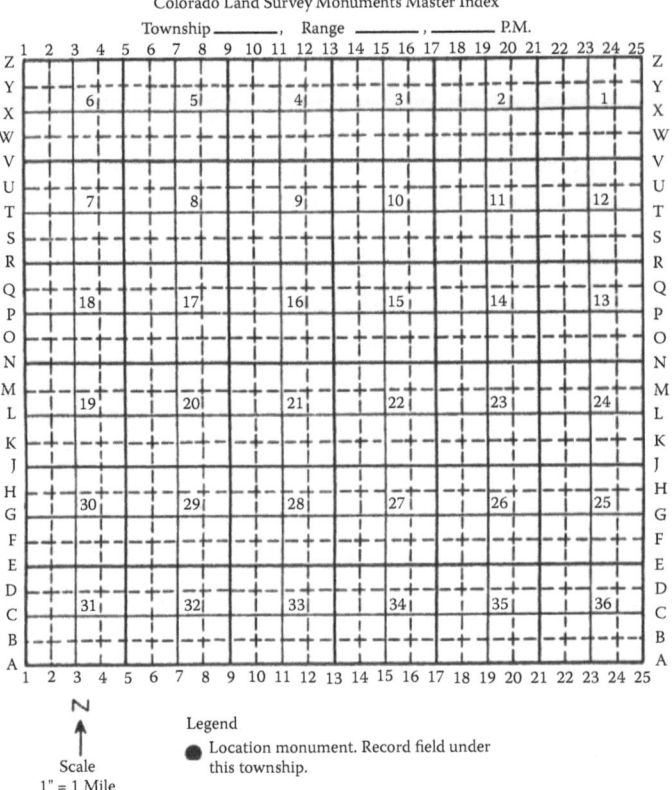

Colorado Land Survey Monuments Master Index

Township _____, Range _____ , _____ P.M.

N
↑
Scale
1" = 1 Mile

Legend
● Location monument. Record field under this township.

EXPLANATION

Each monument record is identified by a numerical-alphabetical index or coordinate number which is related to the township diagram above. For example, the quarter corner common to sections 4 and 5 would have the coordinate number 9-X. See item 7 on monument record.

Monuments which are not located at intersecting lines on the above diagram are identified by reference to the nearest decimal division of the spaces, with the decimal values increasing to the right and upward; e.g., 1.7-C, 8-M.3, 6.2-T.5.

A monument which is common to adjoining townships will be identified by the coordinate in the township in which the record is to be filed, except in the case where a county boundary coincides with a township line, in which case an appropriate coordinate number shall be used for each county involved.

Within each township file, the records shall be inserted in numerical-alphabetical order from top to bottom.

NOTICE

There may be more than one record for a monument. DO NOT REMOVE OLD RECORD when filing new one.

FIGURE 39

Example of Colorado Monument Record, page 2.

Colorado Land Survey Monument Record

Department of Regulatory Agencies
Board of Registration for Professional Engineers and Professional Land Surveyors
1560 Broadway, Suite 1370, Denver, Co 80202
Phone (303) 894-7788 • Fax (303) 894-7790 • TDD Line (303) 894-2900×833

REPORT ONE MONUMENT ONLY ON THIS FORM - REPRODUCTION OF THIS FORM IS AUTHORIZED
All items to be filled in by the Land Surveyor using PERMANENT BLACK LETTERING and lines which can be reproduced

1. TYPE OF MONUMENT: ☐ SECTION CORNER ☐ QUARTER CORNER ☐ BENCH MARK ☐ OTHER _____

2. DESCRIPTION OF MONUMENT FOUND:

3. DESCRIPTION OF MONUMENT ESTABLISHED BY YOU TO PERPETUATE THE LOCATION OF THIS POINT:

3. SKETCH SHOWING RELATIVE LOCATION OF MONUMENT, ACCESSORIES AND REFERENCE POINT STATING WHETHER FOUND OR SET, SHOW SUPPORTING AND/OR CONTRADICTORY EVIDENCE WHERE APPLICABLE:

Stamping on Cap

5. CERTIFICATION

This is to Certify that I was in responsible charge of the surveying work described in this record and that to the best of my knowledge the information presented herein is true and correct.

Name (Print of Type): _____

Firm Name: _____
Firm Address: _____
_____

a. Date of Field Work to Establish, Restore or Rehabilitate
Monument: _____

b. Date Monument was Used as Control: _____

Phone _____

6. LOCATION DIAGRAM
1" = Mile

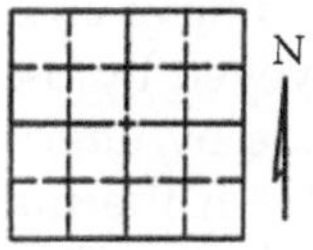

N

————— (Do not fill in) —————

RECEIVED AT OFFICE OF THE COUNTY CLERK

_____ COUNTY

BY: _____

DATE: _____

Record to by filed by Index Reference Number,
Numerically, then, Alphabetically, under
appropriate Township, Range, and Meridian

* = Location of Monument **Signature/date through seal**

7. SEC _____ T _____ R _____ , _____ P.M.
   COUNTY _____ INDEX REF NUMBER _____

**8. SEC _____ T _____ R _____ , _____ P.M.
   County _____ INDEX REF NUMBER _____
   * * To be used only for monuments located on county lines

FIGURE 40
Example of Colorado Monument Record, page 3.

1855 Federal Manual of Surveying Instructions

Below, a partial copy of the 1855 manual is shown (the example of the Field Notes and Plats have been omitted). These pages have been included to provide the following practical use:

- In order to follow in the footsteps of their predecessors, modern surveyors must know the principles, methods, and limits of the original GLO surveys (i.e., the instructions given to the federal surveyors who made the initial surveys).

- As an example, about 80% or roughly 83,000 sq. mi. of Colorado, tied to the 6th Principal Meridian, were initially surveyed under the 1855 *Manual of Surveying Instructions*. Hence, it is very important for anybody who is involved in a resurvey in the area involved to understand the principles and methods used. For example, the 1855 *Manual* calls for standard parallels to be run every 30 miles south of the base line and not every 24 miles as required in later manuals. Another example is the method on how true north was established. Thus, probably 99% of the practicing surveyors of Colorado involved in a resurvey should read and understand this manual. For most areas it is also important to realize that GLO surveying practices have changed with time and while the initial surveys were surveyed by the earlier manual, subsequent surveys were possibly made by later manuals. It should be noted that the manual will set forth the methods and principles of the surveys for a specific time frame. The field notes by contrast will show the numbers, for distances and directions, as well as the type of monuments set, etc.

- On the other hand, the very few people who will be involved in a new GLO survey should read and follow the instructions of the latest (2009) manual, as stipulated by Colorado State Law (CRS 38-550-101 and CRS 38-51-101). As sometimes suggested, it simply does not make sense to read the newest manual in order to find out what was done some 150 years ago.

- Non-Colorado surveyors in other GLO areas can use the concepts shown above by merely substituting the specific date from the manuals published in 1881, 1890, 1894, 1902, 1919, 1930, 1947 or 1973. Probably C. Albert White's *A History of the Rectangular Survey System* should be consulted first.

(This copy of the 1855 manual is taken from an original volume now in the possession of the National Archives.)

INSTRUCTIONS
TO THE
SURVEYORS GENERAL OF PUBLIC
LANDS
OF
THE UNITED STATES,
FOR THOSE
SURVEYING DISTRICTS ESTABLISHED
IN AND SINCE THE YEAR 1850;
CONTAINING, ALSO,
A MANUAL OF INSTRUCTIONS
TO
REGULATE THE FIELD OPERATIONS OF
DEPUTY SURVEYORS,
ILLUSTRATED BY DIAGRAMS.

PRESCRIBED, ACCORDING TO LAW, BY THE PRINCIPAL CLERK OF SURVEYS, PURSUANT TO ORDER OF THE COMMISSIONER OF THE GENERAL LAND OFFICE.

WASHINGTON:
A. O. P. NICHOLSON, PUBLIC PRINTER.
1855.

Editorial Note:

- The "Manual of Surveying Instructions" was published in 1855, 1881, 1890, 1894, 1902, 1919, 1930, 1947, and 1973.

- This 1855 copy of the manual should be consulted for any GLO surveys made between 1855 and 1881, i.e. most of the original GLO surveys in Colorado.

TO THE SURVEYORS GENERAL

OF

PUBLIC LANDS OF THE
UNITED STATES

FOR THE SURVEYING DISTRICTS ESTABLISHED
IN AND SINCE THE YEAR 1850.

By the direction of the COMMISSIONER OF THE GENERAL LAND OFFICE, the accompanying instructions are prescribed for your official government, including a MANUAL OF INSTRUCTIONS to regulate the field operations of your deputy surveyors. The latter is a revision of the Manual of Surveying Instructions prepared for OREGON in 1851, (the edition of which is now exhausted,) and presents, in some respects, more copious illustrations, both in the specimen field notes and in the diagrams, than could be furnished amidst the pressure of the exigency under which the former had to be prepared. It will be observed that, in the former edition, the township and section lines south of the base are made to start therefrom, and close on the first standard parallel south; whereas, under the present instructions, such lines are made to start from the first standard parallel south, and to close to the north on the base: and thus there will be closing corners and starting corners, both on the base and standard lines. Such modification is introduced for the sake of entire uniformity of method in new fields of survey, and will not, of course, affect any past operations under the original instructions.

The starting corners on the base line and on the standards will, of course, be common to two townships or to two sections lying on and north of such lines; and the closing corners on such lines, from the south, should be carefully connected with the former by measurements to be noted in the field book.

Where STONE can be had to perpetuate corner boundaries, such, for obvious reasons, should always be preferred for that purpose, and the dimensions of the stone, as herein prescribed, (on page 9,) are to be regarded as the minimum size; but in localities where it is found practicable to obtain a stone of *increased dimensions,* it is always desirable to do so, particularly for TOWNSHIP CORNERS, and especially for those on base, meridian, and standard lines; and to such purport the deputy surveyor is to be specially instructed.

Prior to entering upon duty, the deputy surveyor is to make himself thoroughly acquainted with the official requirements in regard to field operations in all the details herein set forth, and to be apprized of the weighty moral and legal responsibilities under which he will act.

[iv]

Unfaithfulness in the execution of the public surveys will be detected by special examinations of the work to be made for that purpose, and, when detected, will immediately subject the delinquent deputy and his bondsmen to be sued by the district attorney of the United States, at the instance of the proper surveyor general—the institution of which suit will act at once as a lien upon any property owned by him or them at that time; and such delinquency, moreover, is an offence punishable by the statute, with all the pains and penalties of perjury, (see act of 1846, quoted on pages 19 and 20 hereof,) and will of necessity debar the offending deputy from future employment in like capacity. Hence, in the execution of contracts for surveying public lands, there is every incentive to fidelity that can address itself either to the moral sense, or to motives of private interest.

By order to the Commissioner:

JOHN M. MOORE,
Principal Clerk of Surveys.

GENERAL LAND OFFICE,
February 22, 1855

TABLE OF CONTENTS.

SYSTEM

OF

RECTANGULAR SURVEYING.

1. The public lands of the United States are ordinarily surveyed into rectangular tracts, bounded by lines conforming to the cardinal points.

2. The public lands are laid off, in the first place, into bodies of land of six miles square, called *Townships*, containing as near as may be 23,040 acres. The townships are subdivided into thirty-six tracts called *Sections*, of a mile square, each containing as near as may be, 640 acres. Any number of series of contiguous townships, situate north or south of each other, constitute a *Range*.

The law requires that the lines of the public surveys shall be governed by the true meridian, and that the townships shall be *six miles square*,—two things involving in connexion a mathematical impossibility—for, strictly to conform to the meridian, necessarily throws the township out of square, by reason of the convergency of meridians, and hence, by adhering to the true meridian, results the necessity of departing from the strict requirements of law, as respects the precise area of townships and the subdivisional parts thereof, the township assuming something of a trapezoidal form, which inequality developes itself more and more as such the higher the latitude of the surveys. It is doubtless in view of these circumstances that the law provides (see sec. 2 of the act of May 18, 1796) that the sections of a mile square shall contain the quantity of 640 acres, *as nearly as may be;* and, moreover, provides (see sec. 3 of the act of 10th May, 1800) in the following words: "And in all cases where the exterior lines of the townships, thus to be subdivided into sections or half sections, shall exceed, or shall not extend six miles, the excess or deficiencey shall be specially noted, and added to or deducted from the western or northern ranges of sections or half sections in such township, according as the error may be in running the lines from east to west, or from south to north; the sections and half sections bounded on the northern and western lines of such townships shall be sold as containing only the quantity expressed in the returns and plats, respectively, and all others as containing the complete legal quantity."

[2]

The accompanying diagram, marked A, will serve to illustrate the method of running out the exterior lines of townships, as well on the *north* as on the *south* side of the base line; and the order and mode of subdividing townships will be found illustrated in the accompanying specimen field notes, conforming with the township diagram B. The method here presented is designed to insure as full a compliance with all the requirements, meaning, and intent of the surveying laws as, it is believed, is practicable.

The section lines are surveyed from *south* to north on true meridians, and from *east* to west, in order to throw the excesses or deficiencies in measurements on the north and west sides of the township, as required by law.

3. The townships are to bear numbers in respect to the base line either north or south of it; and the tiers of townships, called "Ranges," will bear numbers in respect to the meridian line according to their relative position to it, either on the east or west.

4. The thirty-six sections into which a township is subdivided are numbered, commencing with number *one* at the *northeast* angle of the township, and proceeding west to number six, and thence proceeding east to number twelve, and so on, alternately, until the number thirty-six in the southeast angle.

5. STANDARD PARALLELS (usually called correction lines) are established at stated intervals to provide for or counteract the error that otherwise would result from the convergency of meridians, and also to arrest error arising from inaccuracies in measurements on meridian lines, which, however, must ever be studiously avoided. On the *north* of the paincipal base line it is proposed to have these standards run at distances of every *four* townships, or twenty-four miles, and on the *south* of the principal base, at distances of every *five* townships, or thirty miles.

OF MEASUREMENTS, CHAINING, AND MARKING.

1. Where uniformity in the variation of the needle is not found, the public surveys must be made with an instrument operating independently of the magnetic needle. Burt's *improved solar compass*, or other istrument of equal utility, must be used of necessity in such cases; and it is deemed best that such instrument should be used under all circumstances. Where the needle can be relied on, however, the ordinary compass may be used in subdividing and meandering.

[3]

2. The township lines, and the subdivision lines, will usually be measured by a two-pole chain of thirty-three feet in length, consisting of fifty links, and each link being seven inches and ninety-two hundredths of an inch long. On uniform and level ground, however, the four-pole chain may be used. Your measurements will, however, always be represented according to the four-pole chain of one hundred links. The deputy surveyor must also have with him a measure of the standard chain, wherewith to compare and adjust the chain in use, from day to day, with punctuality and carefulness; and must return such standard chain to the Surveyor General's office for examination when his work is completed.

OF TALLY PINS.

3. You will use eleven tally pins made of steel, not exceeding fourteen inches in length, weighty enough towards the point to make them drop perpendicularly, and having a ring at the top, in which is to be fixed a piece of red cloth, or something else of conspicuous color, to make them readily seen when stuck in the ground.

PROCESS OF CHAINING.

4. In measuring lines with a two-pole chain, every *five* chains are called "a *tally*," because at that distance the last of the ten tally pins with which the forward chainman set out will have been stuck. He then cries "tally;" which cry is repeated by the other chainman, and each registers the distance by slipping a thimble, button, or ring of leather, or something of the kind, on a belt worn for that purpose, or by some other convenient method. The hind chainman then comes up, and having counted in the presence of his fellow the tally pins which he has taken up, so that both may be assured that none of the pins have been lost, he then takes the forward end of the chain, and proceeds to set the pins. Thus the chainmen alternately chain places, each setting the pins that he has taken up, so that one is forward in all the odd, and the other in all the even tallies. Such procedure, it is believed, tends to insure accuracy in measurement, facilitates the recollection of the distances to objects on the line, and renders a mis-tally almost impossible.

LEVELLING THE CHAIN AND PLUMBING THE PINS.

5. The length of every line you run is to be ascertained by precise horizontal measurement, as nearly approximating to an air line as is pos-

[4]

sible in practice on the earth's surface. This all important object can only be attained by a rigid adherence to the three following observances:

1. Ever keeping the chain *stretched* to its utmost degree of tension on even ground.

2. On uneven ground, keeping the chain not only stretched as aforesaid, but horizontally *levelled*. And when ascending and descending steep ground, hills, or mountains, the chain will have to be *shortened* to one-half its length, (and sometimes more,) in order accurately to obtain the true horizontal measure.

3. The careful plumbing of the tally pins, so as to attain precisely *the spot* where they should be stuck. The more uneven the surface, the greater the caution needed to set the pins.

MARKING LINES.

6. All lines on which are to be established the legal corner boundaries are to be marked after this method, viz: Those trees which may intercept your line must have two chops or notches cut on each side of them without any other marks whatever. These are called *"sight trees," "line trees,"* or *"station trees."*

A sufficient number of other trees standing nearest to your line, on either side of it, are to be *blazed* on two sides diagonally, or quartering towards the line, in order to render the line conspicuous, and readily to be traced, the blazes to be opposite each other, coinciding in direction with the line where the trees stand very near it, and to approach nearer each other the further the line passes from the blazed trees. Due care must ever be taken to have the lines so well marked as to be readily followed.

ON TRIAL, OR RANDOM LINES,

the trees are not to be blazed, unless occasionally from indispensable necessity, and then it must be done so guardedly as to prevent the possibility of confounding the marks of the trial line with the *true*. But bushes and limbs of trees may be lopped, and *stakes set* on the trial, or random line, at every ten chains, to enable the surveyor on his return to follow and correct the trial line, and establish therefrom the *true line*. To prevent confusion, the temporary stakes set on the trial, or random lines, must be *pulled up* when the surveyor returns to establish the true line.

[5]

INSUPERABLE OBJECTS ON LINE—WITNESS POINTS.

7. Under circumstances where your course is obstructed by impassable obstacles, such as ponds, swamps, marshes, lakes, rivers, creeks, &c., you will prolong the line across such obstacles by taking the necessary right angle offsets; or, if such be inconvenient, by a traverse or trigonometrical operation, until you regain the line on the opposite side. And in case a north and south, or a true east and west, line is regained in advance of any such obstacle, you will prolong and mark the line back to the obstacle so passed, and state all the particulars in relation thereto in your field book. And at the intersection of lines with both margins of impassable obstacles, you will establish a *Witness Point*, (for the purpose of perpetuating the intersections therewith,) by setting a post, and giving in your field book the course and distance therefrom to two trees on opposite sides of the line, each of which trees you will mark with a blaze and notch facing the post; but on the margins of navigable water courses, or navigable lakes, you will mark the trees with the proper number of the fractional section, township, and range.

The best marking tools adapted to the purpose must be provided for marking neatly and *distinctly* all the letters and figures required to be made at corners; and the deputy is to have always at hand the necessary implements for keeping his marking irons in order; for which purpose a rat-tail file and a small whetstone will be found indispensable.

ESTABLISHING CORNER BOUNDARIES.

To procure the faithful execution of this portion of a surveyor's duty is a matter of the utmost importance. After a true coursing, and most exact measurements, the corner boundary is the consummation of the work, for which all the previous pains and expenditures have been incurred. If, therefore, the corner boundary be not perpetuated in a permanent and workmanlike manner, the *great aim* of the surveying service will not have been attained. A boundary corner, in a timbered country, is to be a *tree*, if one be found at the precise spot; and if not, *a post* is to be planted thereat; and

the position of the corner post is to be indicated by trees adjacent, the angular bearings and distances of which from the corner are facts to be ascertained and registered in your field book. (See article, "Bearing trees.")

[6]

In a region where stone abounds the corner boundary will be a small *monument of stones* along side of a single marked stone for a township corner, and a single stone for all other corners.

In a region where timber is not near, and stone not found, the corner will be a *mound* of earth, of prescribed size, varying to suit the case.

The following are the different points for perpetuating corners, viz:

1. For township boundaries, at intervals of every six miles.

2. For section boundaries, at intervals of every mile, or 80 chains.

3. For quarter section boundaries, at intervals of every half mile, or 40 chains. Exceptions, however, occur on east and west lines, as explained hereafter.

[The half quarter section boundary is not marked in the field, but is regarded by the law as a point intermediate between the half mile or quarter section corners. See act of 24th April, 1820, entitled "An act making further provision for the sale of the public lands," which act refers to the act of Congress passed on the 11th of February, 1805, entitled "An act concerning the mode of surveying the public lands of the United States," for the manner of ascertaining the corners and contents of half quarter sections.][1]

4. MEANDER CORNER POSTS are planted at all those points where the township or section lines intersect the banks of such rivers, bayous, lakes, or islands, as are by law directed to be meandered.

The courses and distances on meandered navigable streams govern the calculations wherefrom are ascertained the true areas of the tracts of land (sections, quarter sections, &c.) known to the law as *fractional*, and binding on such streams.

MANNER OF ESTABLISHING CORNERS BY MEANS OF POSTS.

Township, sectional, or mile corners, and quarter sectional or half mile corners, will be perpetuated by planting a post at the place of the corner, to be formed of the most durable wood of the forest at hand.

The posts must be set in the earth by digging a hole to admit them *two feet* deep, and must be very securely rammed in with earth, and also with stone, if any be found at hand. The portion of the post which protrudes above the earth must be *squared* off sufficiently smooth to admit of receiving the marks thereon, to be made with appropriate marking irons, indicating what it stands for. Thus the sides of *township*

---

1. The subdivision of the half-quarter section into quarter-quarter sections is authorised by "An act supplementary to the several laws for the sale of the public lands," approved April 5, 1832.

[7]

corner posts should square at least *four* inches, (the post itself being *five* inches in diameter,) and must protrude *two feet* at least above the ground; the sides of *section corner posts* must square at least *three inches*, (the post itself being four inches in diameter,) and protrude *two feet* from the ground; and the *quarter section corner posts* and *meander corner posts* must be *three* inches *wide*, presenting *flattened* surfaces, and protruding *two* feet from the ground.

Where a township post is a corner common to four townships, it is to be set in the earth *diagonally*, thus:

N

W ◇ E

S

On each surface of the post is to be marked the number of the particular township, and its range, which it *faces*. Thus, if the post be a common boundary to four townships—say *one* and *two*, south of the base line, of range *one*, west of the meridian; also to townships *one* and *two*, south of the base line of range *two*, west of the meridian, it is to be marked thus:

$$\text{From N. to E.} \left\{ \begin{matrix} \text{R.} & 1 & \text{W.} \\ \text{T.} & 1 & \text{S.} \\ \text{S.} & 31 & \end{matrix} \right\} \qquad \text{from E. to S.} \left\{ \begin{matrix} 1 & \text{W.} \\ 2 & \text{S.} \\ 6 & \end{matrix} \right\}$$

$$\text{From N. to W.} \left\{ \begin{matrix} 2 & \text{W.} \\ 1 & \text{S.} \\ 36 & \end{matrix} \right\} \qquad \text{from W. to S.} \left\{ \begin{matrix} 2 & \text{W.} \\ 2 & \text{S.} \\ 1 & \end{matrix} \right\}$$

These marks are not only to be distinctly but *neatly* cut into the wood, at least the eighth of an inch deep; and to make them yet more *conspicuous* to the eye of the anxious explorer, the deputy must apply to all of them a *streak of red chalk*.

Section or mile posts, being corners of sections, and where such are common to *four* sections, are to be set *diagonally* in the earth, (in the manner provided for township corner posts;) and on each side of the squared surfaces (made smooth, as aforesaid, to receive the marks) is to be marked the appropriate *number* of the particular one of the *four sections*, respectively, which such side *faces*; also on one side thereof are to be *marked* the numbers of its *township* and *range*; and to make such marks yet more *conspicuous*, in manner aforesaid, a streak of *red chalk* is to be applied.

In every township, subdivided into thirty-five sections, there are twenty-five interior section corners, each of which will be *common to four sections*.

A quarter section, or half mile post, is to have no other mark on it than ¼ S., to indicate what it stands for.

[8]

NOTCHING CORNER POSTS.

Township corner posts, common to four townships, are to be notched with *six* notches on each of the four angles of the squared part set to the cardinal points.

All mile posts *on township lines* must have as many notches on them, on two opposite *angles* thereof, as they are miles distant from the township corners, respectively. Each of the posts at the corners of sections in the *interior* of a township must indicate, by a number of notches on each of its four

corners directed to the cardinal points, the corresponding number of miles that it stands from the *outlines* of the township. The four sides of the post will indicate the number of the section they respectively *face*. Should a tree be found at the place of any corner, it will be marked and notched as aforesaid, and answer for the corner in lieu of a post, the kind of tree and its diameter being given in the field notes.

BEARING TREES.

The position of all corner posts, or corner trees, of whatever description, that may be established, is to be evidenced in the following manner, viz: From such post or tree the courses must be taken and the distances measured to two or more adjacent trees in opposite directions, as nearly as may be, and these are called "bearing trees." Such are to be distinguished by a large *smooth blaze*, with a *notch* at its lower end, facing the corner, and in the blaze is to be marked the number of the *range, township,* and *section*; but at quarter section corners nothing but ¼ S. need be marked. The letters B. T. (bearing tree) are also to be marked upon a smaller blaze directly under the large one, and as near the ground as practicable.

At all township corners, and at all section corners, on range or township lines, *four* bearing trees are to be marked in this manner, one in each of the adjoining sections.

At interior section corners *four* trees, one to stand within each of the four sections to which such corner is common, are to be marked in manner aforesaid, if such be found.

A tree supplying the place of a corner post is to be marked in the manner directed for posts; but if such tree should be a beech, or other *smooth bark* tree, the marks may be made on the *bark*, and the tree notched.

From quarter section and meander corners two bearing trees are to be marked, one within each of the adjoining sections.

[9]

Where the requisite number of "bearing trees" is not to be found at convenient and suitable distances, such as are found are to be marked as herein directed; but in all cases of deficiency in the number of bearing tree, (unless, indeed, the boundary itself be a tree,) a *quadrangular trench*, with sides of *five* feet, and with the angles to the cardinal points, must be spaded up outside the corner, as a centre, and the earth carefully thrown on the inside, so as to form a range of earth, which will become covered with grass, and present a small square elevation, which in aftertime will serve to mark, unmistakably, the spot of the corner.

CORNER STONES.

Where it is deemed best to use STONES for boundaries, in lieu of posts, you may, at *any* corner, insert endwise into the ground, to the depth of 7 or 8 inches, a stone, the number of cubic inches in which shall not be less than the number contained in a stone 14 inches long, 12 inches wide, and 3 inches thick—equal to 504 cubic inches—the edges of which must be set north and south, on north and south lines, and east and west, on east and west lines; the dimensions of each stone to be given in the field notes at the time of establishing the corner. The kind of stone should also be stated.

MARKING CORNER STONES.

Stones at township corners, common to four townships, must have *six* notches, cut with a pick or chisel on each edge or side towards the cardinal points; and where used as section corners on the range and township lines, or as section corners in the interior of a township, they will also be notched, to correspond with the directions given for notching posts similarly situated.

Posts or stones at township corners on the base and standard lines, and which are common to two townships on the north side thereof, will have *six* notches on each of the *west*, *north*, and *east* sides or edges; and where such stones or posts are set for corners to two townships *south* of the base or standard, *six* notches will be cut on each of the west, *south*, and east sides or edges.

Stones, when used for quarter section corners, will have ¼ cut on them—on the west side on north and south lines, and on the north side on east and west lines.

[10]

MOUNDS.

Whenever bearing trees are not found, mounds of earth, or stone, are to be raised *around posts* on which the corners are to be marked in the manner aforesaid. Wherever a mound of earth is adopted, the same will present a conical shape; but at its base, on the earth's surface, a *quadrangular trench* will be dug; by the "trench" (here meant) is to be understood a *spade deep* of earth thrown up from the four sides of the line, *outside* the trench, so as to form a *continuous elevation along its outer edge*. In mounds of earth, common to *four* townships or to *four* sections, they will present the *angles* of the quadrangular trench *(diagonally)* towards the cardinal points. In mounds, common only to *two* townships or *two* sections, the *sides* of the quadrangular trench will *face* the cardinal points. The sides of the quadrangular trench at the base of a township mound are to be *six* feet, the height of mound *three* feet.

At section, quarter section, and meander corners, the sides of the quadrangular trench at base of mounds are to be *five* feet, and the conical height *two and a half feet*.

Prior to piling up the earth to construct a mound, there is to be dug a spadefull or two of earth from the corner boundary point, and in the cavity so formed is to be deposited a *marked stone*, or a portion of *charcoal*, (the quantity whereof is to be noted in the field book;) or in lieu of charcoal or marked stone, a *charred stake* is to be driven twelve inches down into such centre point: either of those will be a *witness* for the future, and whichever is adopted, the fact is to be noted in the field book.

When mounds are formed of *earth*, the spot from which the earth is taken is called the *"pit,"* the centre of which ought to be, wherever practicable, at a uniform distance and in a uniform direction from the centre of the mound. There is to be a "pit" on *each* side of every mound, distant eighteen inches outside of the trench. The trench may be expected hereafter to be covered by tufts of grass, and thus to indicate the place of the mound, when the mound itself may have become obliterated by time or accident.

At meander corners the "pit" is to be directly on the line, *eight links* further from the water than the mound. Wherever necessity is found for deviating from these rules in respect to the "pits," the course and distance to each is to be stated in the field books

Perpetuity in the mound is a great desideratum. In forming it with light alluvial soil the surveyor may find it necessary to make due allowance for the future settling of the earth, and thus making the mound

[11]

more elevated than would be necessary in a more compact and tenacious soil, and increasing the base of it. In so doing, the relative proportions between the township mound and other mounds is to be preserved as nearly as may be.

The earth is to be pressed down with the shovel during the process of piling it up. Mounds are to be *covered* with sod, grass side up, where sod is to be had: but, in forming a mound, *sod* is NEVER to be *wrought up* with the earth, because sod decays, and in the process of decomposing it will cause the mound to become porous, and therefore liable to premature destruction.

POSTS IN MOUNDS

must show above the top of the mound ten or twelve inches, and be notched and marked precisely as they would be for the same corner without the mound.

MOUND MEMORIALS.

Besides the *charcoal*, marked *stone* or *charred stake*, one or the other of which must be lodged in the earth at the point of the corner, the deputy surveyor is recommended to plant *midway* between each pit and the trench, seeds of some tree, (those of fruit trees adapted to the climate being always to be preferred,) so that, in course of time, should such take root, a small clump of trees may possibly hereafter note the place of the corner. The facts of planting such seed, and the kind thereof, are matters to be truthfully noted in the field book.

WITNESS MOUNDS TO TOWNSHIP OR SECTION CORNERS.

If a township or section corner, in a situation here bearing or witness trees are not found within a reasonable distance therefrom, shall fall within a ravine, or in any other situation where the nature of the ground, or the circumstances of its locality, shall be such as may prevent, or prove unfavorable to, the erection of a mound, you will perpetuate such corner by selecting in the immediate vicinity thereof a suitable plot of ground as a site for a bearing or *witness mound*, and erect thereon a mound of earth in the same manner and conditioned in every respect, with *charcoal*, *stone*, or *charred stake* deposited beneath, as before directed; and measure and state in your field book the distance and course from the position of the true corner of the bearing or witness mound so placed and erected.

[12]

DOUBLE CORNERS.

Such corners are to be nowhere except on the base and standard lines, whereon are to appear both the corners which mark the intersections of the lines which close thereon, and those from which the surveys start on the north. On these lines, and at the time of running the same, the township, section, and quarter section corners are to be planted, and each of these is a corner commmon to *two*, (whether township or section corners,) on the north side of the line, and must be so marked.

The corners which are established on the standard parallel, at the time of running it, are to be known as *"standard corners,"* and, in addition to all the *ordinary* marks, (as herein prescribed,) they will be marked with the letters S. C. Closing corners will be marked with the letters C. C. in addition to other marks.

The standard parallels are designed to be run in *advance* of the contiguous surveys on the south of them, but circumstances may exist which will *impede* or temporarily delay the *due* extension of the standard; and when, from uncontrollable causes, the *contiguous townships* must be surveyed in advance of the time of extending the standard, in any such event it will become the duty of the deputy who shall afterwards survey any such standard to plant thereon the *double set* of corners, to wit, the standard corners, to be marked S. C., and the closing ones which are to be marked C. C.; and to make such measurements as may be necessary to connect the closing corners and complete the unfinished meridianal lines of such contiguous and prior surveys, on the principles herein set forth, under the different heads of "exterior or township lines," and of "diagram B."

You will recollect that the corners, (whether township or section corners,) which are *common to two*, (two townships or two sections,) are not to be planted *diagonally* like those which are common to *four*, but with the flat sides facing the cardinal points, and on which the marks and notches are made as usual. This, it will be perceived, will serve yet more fully to distinguish the standard parallels from all other lines.

THE MEANDERING OF NAVIGABLE STREAMS.

1. Standing with the face looking *down* stream, the bank on the *left* hand is termed the "left bank," and that on the *right* hand the "right bank." These terms are to be universally used to distinguish the two banks of a river or stream.

[13]

2. Both banks of *navigable* rivers are to be meandered by taking the courses and distances of their sinuosities, and the same are to be entered in the field book.

At those points where either the township or section lines intersect the banks of a navigable stream, POSTS, or, where necessary, MOUNDS of *earth* or *stone*, are to be established at the time of running these lines. These are called "meander corners;" and in meandering you are to commence at one of those corners on the township line, coursing the banks, and measuring the distance of each course from your commencing corner to the next "meander corner," upon the same or another boundary of the same township, carefully noting your intersection with all intermediate meander corners. By the same method you are to meander the opposite bank of the same river.

The crossing distance *between* the MEANDER CORNERS on same line is to be ascertained by triangulation, in order that the river may be protracted with entire accuracy. The particulars to be given in the field notes.

3. You are also to meander, in manner aforesaid, all *lakes* and deep ponds of the area of twenty-five acres and upwards; also navigable bayous; *shallow* ponds, readily to be drained, or likely to dry up, are not to be meandered.

You will notice all streams of water falling into the river, lake, or bayou you are surveying, stating the width of the same at their mouth; also all springs, noting the size thereof and depth, and whether the water be pure or mineral; also the head and mouth of all bayous; and all islands, rapids, and bars are to be noticed, with intersections to their upper and lower points to establish their exact situation. You will also note the elevation of the banks of rivers and streams, the heights of falls and cascades, and the length of rapids.

4. The precise relative position of islands, in a township made fractional by the river in which the same are situated, is to be determined trigonometrically—sighting to a flag or other fixed object on the island, from a special and carefully measured base line, connected with the surveyed lines, on or near the river bank, you are to form connexion between the meander corners on the river to points corresponding thereto, in direct line, on the bank of the island, and there establish the proper meander corners, and calculate the distance across.

5. In meandering lakes, ponds, or bayous, you are to commence at a meander corner upon the township line, and proceed as above directed for the banks of a navigable stream. But where a lake, pond, or bayou

[14]

lies entirely within the township boundaries, you will commence at a meander corner established in subdividing, and from thence take the courses and distances of the entire margin of the same, noting the intersection with all the meander corners previously established thereon.

6. To meander a pond lying entirely within the boundaries of a section, you will run and measure *two* lines thereunto from the nearest section or quarter section corner on *opposite* sides of such pond, giving the courses of such lines. At *each* of the points where such lines shall intersect the margin of such pond, you will establish a *witness* point, by fixing a post in the ground, and taking bearings to any adjacent trees, or, if necessary, raising a mound.

The relative position of these points being thus definitely fixed in the section, the meandering will commence at one of them, and be continued to the other, noting the intersection, and thence to the beginning. The proceedings are to be fully entered in the field book.

7. In taking the connexion of an island with the main land,

when there is no meander corner in line, opposite thereto, to sight from, you will measure a special base from the meander corner nearest to such island, and from such base you will triangulate to some fixed point on the shore of the island, ascertain the distance across, and there establish a *special* meander corner, wherefrom you will commence to meander the island.

The field notes of meanders will be set forth in the body of the field book according to the dates when the work is performed, as illustrated in the specimen notes annexed. They are to state and describe particularly the meander corner from which they commenced, each one with which they close, and are to exhibit the meanders of each fractional section separately; following, and composing a part of such notes, will be given a description of the land, timber, depth of inundation to which the bottom is subject, and the banks, current, and bottom of the stream or body of water you are meandering.

9. No blazes or marks of any description are to be made on the lines meandered between the established corners, but the utmost care must be taken to pass no object of topography, or *change* therein, without giving a particular description thereof in its proper place in your meander notes.

[15]

OF FIELD BOOKS.

The FIELD NOTES afford the elements from which the plats and calculations in relation to the public surveys are made. They are the source wherefrom the description and evidence of locations and boundaries are officially delineated and set forth. They therefore must be a faithful, distinct and minute record of every thing officially done and observed by the surveyor and his assistants, pursuant to instructions, in relation to running, measuring, and marking lines, establishing boundary corners, &c.; and present, as far as possible, a full and complete *topographical description* of the country surveyed, as to every matter of useful information, or likely to gratify public curiosity.

There will be sundry separate and distinct field books of surveys, as follows:

Field notes of the MERIDIAN and BASE lines, showing the establishment of the *township, section* or mile, and *quarter section* or half mile, boundary corners thereon; with the crossings of streams, ravines, hills, and mountains; character of soil, timber, minerals, &c.

Field notes of the "STANDARD PARALLELS, or correction lines," will show the establishment of the township, section, and quarter section corners, besides exhibiting the topography of the country on line, as required on the base and meridian lines.

Field notes of the EXTERIOR lines of TOWNSHIPS, showing the establishment of corners on lines, and the topography, as aforesaid.

Field notes of the SUBDIVISIONS OF TOWNSHIPS, into sections and quarter sections.

The field notes must in all cases be taken precisely in the order in which the work is done on the ground, and the date of each day's work must follow immediately after the notes

thereof. The *variation of the needle* must always occupy a *separate line* preceding the notes of measurements on line.

The exhibition of every mile of surveying, whether on township or subdivisional lines, must be *complete in itself*, and be separated by a black line drawn across the paper.

The description of the surface, soil, minerals, timber, undergrowth, &c., on *each mile* of line is to follow the notes of survey of such line, and not be mixed up with them.

No abbreviations of words are allowable, except of such words as are *constantly* occurring, such as *"sec."* for *"section;" "in. diam,"* for

[16]

"inches diameter;" "chs." for *"chains;" "lks."* for *"links;" "dist."* for *"distant,"* &c. Proper names must never be abbreviated, however often their recurrence.

The nature of the subject-matter of the field book is to form its title page, showing the State or Territory where such survey lies, by whom surveyed, and the dates of commencement and completion of the work. The second page is to contain the names and duties of assistants. Whenever a new assistant is employed, or the duties of any one of them are changed, such facts, with the reasons therefor, are to be stated in an appropriate entry immediately preceding the notes taken under such changed arrangements. With the notes of the *exterior* lines of townships, the deputy is to submit a plat of the lines run, on a scale of two inches to the mile, on which are to be noted all the objects of topography on line necessary to illustrate the notes, viz: the distances on line at the crossings of streams, so far as such can be noted on the paper, and the direction of each by an arrow-head pointing down stream; also the intersection of line by prairies, marshes, swamps, ravines, ponds, lakes, hills, mountains, and all other matters indicated by the notes, to the fullest extent practicable.

With the instructions for making subdivisional surveys of townships into sections, the deputy will be furnished by the Surveyor General with a diagram of the *exterior* lines of the townships to be subdivided, (on the above named scale,) upon which are carefully to be laid down the measurements of each of the section lines on such boundaries whereon he is to close, the magnetic variation of each mile, and the particular description of each corner. P. in M. signifies post in mound. And on such diagram the deputy who subdivides will make appropriate sketches of the various objects of topography as they occur on his lines, so as to exhibit not only the points on line at which the same occur, but also the direction and position of each between the lines, or within each section, so that every object of topography may be properly completed or connected in the showing.

These notes must be distinctly written out, in language precise and clear, and their figures, letters, words, and meaning are always to be unmistakable. No leaf is to be cut or mutilated, and none to be taken out, whereby suspicion might be created that the missing leaf contained matter which the deputy believed it to be his interest to conceal.

[17]

SUMMARY OF OBJECTS AND DATA REQUIRED TO BE NOTED.

1. The precise length of every line run, noting all necessary offsets therefrom, with the reason and mode thereof.

2. The kind and diameter of all *"bearing trees,"* with the course and distance of the same from their respective corners; and the precise relative position of WITNESS CORNERS to the *true corners.*

3. The kind of materials (earth or stone) of which MOUNDS are constructed—the fact of their being conditioned according to instructions—with the course and distance of the *"pits,"* from the centre of the mound, where necessity exists for deviating from the *general* rule.

4. *Trees on line.* The name, diameter, and distance on line to all trees which it intersects.

5. Intersections by line of *land objects.* The distance at which the line first intersects and then leaves every *settler's claim and improvement*; prairie; river, creek, or other "bottom;" or swamp, marsh, grove, and wind fall, with the course of the same at both points of intersection; also the distances at which you begin to ascend, arrive at the top, begin to descend, and reach the foot of all remarkable hills and ridges, with their courses, and *estimated* height, in feet, above the level land of the surrounding country, or above the bottom lands, ravines, or waters near which they are situated.

6. Intersections by line of *water objects.* All rivers, creeks, and smaller streams of water which the line crosses; the distance on line at the points of intersection, and their *widths on line.* In cases of *navigable* streams, their width will be ascertained between the *meander corners,* as set forth under the proper head.

7. The land's *surface*—whether level, rolling, broken, or hilly.

8. The *soil*—whether first, second, or third rate.

9. *Timber*—the several kinds of timber and undergrowth, in the order in which they predominate.

10. *Bottom lands*—to be described as wet or dry, and if subject to inundation, state to what depth.

11. *Springs of water*—whether fresh, saline, or mineral, with the course of the stream flowing from them.

12. *Lakes and ponds*—describing their banks and giving their height, and also the depth of water, and whether it be pure or stagnant.

13. *Improvements.* Towns and villages; Indian towns and wigwams; houses or cabins; fields, or other improvements; sugar tree groves, sugar camps, mill seats, forges, and factories.

[18]

14. *Coal* banks or beds; *peat* or turf grounds; *minerals* and ores; with particular description of the same as to quality and extent, and all *diggings* therefor; also *salt* springs and licks. All reliable information you can obtain respecting these objects, whether they be on your immediate line or not, is to appear in the general description to be given at the end of the notes.

15. *Roads and trails,* with their directions, whence and whither.

16. Rapids, cataracts, cascades, or falls of water, with the height of their fall in feet.

17. Precipices, caves, sink-holes, ravines, stone quarries, ledges of rocks, with the kind of stone they afford.

18. *Natural curiosities,* interesting fossils, petrifactions, organic remains, &c.; also all ancient works of art, such as mounds, fortifications, embankments, ditches, or objects of like nature.

19. The *variation* of the needle must be noted at all points or places on the lines where there is found any material *change* of variation, and the position of such points must be perfectly identified in the notes.

20. Besides the ordinary notes taken on line, (and which must always be written down on the spot, leaving nothing to be supplied by memory,) the deputy will subjoin, at the conclusion of his book, such further description or information touching any matter or thing connected with the township (or other survey) which he may be able to afford, and may deem useful or necessary to be known—with a *general description* of the township in the *aggregate,* as respects the face of the country, its soil and geological features, timber, minerals, waters, &c.

SWAMP LANDS.

By the act of Congress approved September 28, 1850, swamp and overflowed lands "unfit for cultivation" are granted to the State in which they are situated. In order clearly to define the quantity and locality of such lands, the field notes of surveys, in addition to the other objects of topography required to be noted, are to indicate the points at which you enter all lands which are evidently subject to such grant, and to show the distinctive character of the land so noted; whether it is a swamp or marsh, or otherwise subject to inundation to an extent that, without artificial means, would render it "unfit for cultivation." The depth of inundation is to be stated, as determined from indications on the trees where timber exists; and its frequency is to be set forth as accurately as may be, either from your own knowledge of the general

[19]

character of the stream which overflows, or from reliable information to be obtained from others. The words "unfit for cultivation," are to be employed in addition to the usual phraseology in regard to entering or leaving such swamps, marshy, or overflowed lands. It may be that sometimes the margin of bottom, swamp, or marsh, in which such uncultivable land exists, is not identical with the margin of the body of land "unfit for cultivation;" and in such cases a separate entry must be made for each opposite the marginal distance at which they respectively occur.

But in cases where lands are overflowed by *artificial* means, (say by dams for milling, logging, or for other purposes,) you are not officially to regard such overflow, but will continue your lines across the same without setting meander posts, stating particularly in the notes the depth of the water, and how the overflow was caused.

SPECIAL INSTRUCTION RESPECTING THE NOTING OF SETTLERS' CLAIMS IN OREGON, WASHINGTON, AND NEW MEXICO.

The law requires that such claims should be laid down temporarily on the township plats; in order to do which, it is indispensably necessary to obtain, to some extent, connexions of these claims with the lines of survey. Under the head of "intersection by line of land objects," the deputy is required to note the *points* in line *whereat* it may be intersected by such claims; but, in addition thereto, there must be obtained at least *one angle* of each claim, with its course and distance either from the point of intersection, or from an established corner boundary, so that its connexion with the regular survey will be legally determined. If the settler's *dwelling* or barn is visible from line, the bearings thereof should be carefully taken from *two* points noted on line, and set forth in the field notes.

AFFIDAVITS TO FIELD NOTES.

At the close of the notes and the *general description* is to follow an affidavit, a form for which is given; and to enable the deputy surveyor fully to understand and appreciate the responsibility under which he is acting, his attention is invited to the provisions of the second section of the act of Congress, approved August 8th, 1846, entitled "An act to equalize the compensation of the surveyors general of the public lands of the United States, and for other purposes," and which is as follows:

"Sec. 2. That the surveyors general of the public lands of the United

[20]

States, in addition to the oath now authorized by law to be administered to deputies on their appointment to office, shall require each of their deputies, on the return of his surveys, to take and subscribe an oath or affirmation that those surveys have been faithfully and correctly executed according to law and the instructions of the surveyor general; and on satisfactory evidence being presented to any court of competent jurisdiction, that such surveys, or any part thereof, had not been thus executed, the deputy making such false oath or affirmation shall be deemed guilty of perjury, and shall suffer all the pains and penalties attached to that offence; and the district attorney of the United States for the time being, in whose district any such false, erroneous, or fraudulent surveys shall have been executed, shall, upon the application of the proper surveyor general, immediately institute suit upon the bond of such deputy; and the institution of such suit shall act as a lien upon any property owned or held by such deputy, or his sureties, at the time such suit was instituted."

Following the "general description" of the township is to be "A list of the names of the individuals employed to assist in running, measuring and marking the lines and corners described in the foregoing field notes of township No. _____ of the BASE LINE of range No. _____ of the _____ MERIDIAN, showing the respective capacities in which they acted."

FORM OF OFFICIAL OATHS TO BE TAKEN PRIOR TO ENTERING UPON DUTY.

For a deputy surveyor.

I, A. B., having been appointed a deputy urveyor of the lands of the United States in _____, do solemnly swear (or affirm) that I will well and faithfully, and to the best of my skill and ability, execute the duties confided to me pursuant to a contract with C. D., surveyor general of public lands in _____, bearing date the _____ day of _____, 18 , according to the laws of the United States and the instructions received from the said surveyor general.

(To be sworn and subscribed before a justice of the peace, or other officer authorized to administer oaths.)

For chainman.

I, E. F., do solemnly swear (or affirm) that I will faithfully execute the duties of chain carrier; that I will leel the chain upon uneven ground, and plumb the tally pins, whether by sticking or dropping the

[21]

same; that I will report the true distance to all notable objects, and the true length of all lines that I assist in measuring, to the best of my skill and ability.

(To be sworn and subscribed as above.)

For flagman or axeman.

I, G. H., do solemnly swear (or affirm) that I will well and truly perform the duties of _____, according to instructions given me, and to the best of my skill and ability.

(To be sworn and subscribed as above.)

EXTERIORS OR TOWNSHIP LINES.

The principal meridian, the base line, and the standard parallels having been first run, measured, and marked, and the corner boundaries thereon established, according to instructions, the process of running, measuring, and marking the exterior lines of townships will be as follows:

Townships situated NORTH of the base line, and WEST of the principal meridian.

Commence at No. 1, (see figures on diagram A,) being the southwest corner of T. 1 N—.R. 1 W., as established on the base line; thence north, on a true meridian line, four hundred and eighty chains, establishing the section and quarter section corners thereon, as per instructions, to No. 2, whereat establish the corner of Tps. 1 and 2 N—.Rs. 1 and 2 W.; thence east, on a random or trial line, setting *temporary* section and quarter section stakes, to No. 3, where measure and note the distance at which the line intersects the eastern boundary, north or south of the *true* or established corner. Run and measure westward, on the true line, (taking care to note all the land and water crossings, &c., as per instructions,) to No.

4, which is identical with No. 2, establishing the section and quarter section PERMANENT CORNERS on said line. Should it happen, however, that such random line falls short, or overruns in length, or intersects the eastern boundary of the township at more than three chains and fifty links distance from the *true* corner thereon, as compared with the corresponding boundary on the south, (either of which would indicate an important error in the surveying,) the lines must be *retraced*, even if found necessary to remeasure the meridianal

[22]

boundaries of the township, (especially the western boundary,) so as to discover and correct the error; in doing which, the *true* corners must be established and marked, and the *false ones* destroyed and obliterated, to prevent confusion in future; and *all the facts* must be distinctly set forth in the notes. Thence proceed in a similar manner from No. 4 to No. 5, No. 5 to No. 6, No. 6 to No. 7, and so on to No. 10, the southwest corner of T. 4 N—.R. 1 W. Thence north, still on a true meridian line, establishing the mile and half-mile corners, until reaching the STANDARD PARALLEL or correction line; throwing the *excess* over, or *deficiency* under, *four hundred and eighty chains*, on the *last* half-mile, according to law, and at the intersection establishing the "CLOSING CORNER," the distance of which *from* the standard corner must be measured and noted as required by the instructions. But should it ever so happen that some impassable barrier will have prevented or delayed the extension of the standard parallel along and above the field of present survey, then the deputy will plant, in place, the corner for the township, subject to correction thereafter, should such parallel be extended.

NORTH of the base line, and EAST of the
principal meridian.

Commence at No. 1, being the *southeast* corner of T. 1 N—.R. 1 E., and proceed as with townships situated "north and west," except that the *random* or trial lines will be run and measured *west*, and the *true* lines east, throwing the excess over or deficiency under four hundred and eighty chains on the *west end* of the line, as required by law; wherefore the surveyor will commence his measurement with the length of the deficient or excessive half section boundary on the west of the township, and thus the remaining measurements will all be *even* miles and half-miles.

METHOD OF SUBDIVIDING.

1. The first mile, both of the south and east boundaries of each township you are required to subdivide, is to be carefully traced and measured before you enter upon the subdivision thereof. This will enable you to observe any change that may have taken place in the magnetic variation, as it existed at the time of running the township lines, and will also enable you to compare your chaining with that upon the township lines.

2. Any discrepancy, arising either from a change in the magnetic variation or a difference in measurement, is to be carefully noted in the field notes.

[23]

3. After adjusting your compass to a variation which you have thus found will retrace the eastern boundary of the township, you will commence at the corner to sections 35 and 36, on the south boundary, and run a line due north, forty chains, to the quarter section corner which you are to establish between sections 35 and 36; continuing due north forth chains further, you will establish the corner to sections 25, 26, 35 and 36.

4. From the section corner last named, run a *random* line, without blazing, *due east*, for corner of sections 25 and 36, in east boundary, and at forty chains from the starting point set a post for *temporary* quarter section corner. If you intersect exactly at the corner, you will blaze your random line back, and establish it as the *true* line; but if your random line intersects the said east boundary, either north or south of said corner, you will measure the distance of such intersection, from whih you will calculate a course that will run a *true* line back to the corner from which your random started. You will establish the *permanent* quarter section corner at a point equidistant from the two terminations of the *true* line.

5. From the corner of sections 25, 26, 35, 36, run due north between sections 25 and 26, setting the quarter section post, as before, at forty chains, and at eighty chains establishing the corner of sections 23, 24, 25, 26. Then run a *random due east* for the corner of sections 24 and 25 in east boundary; setting temporary quarter section post at forty chains; correcting back, and establishing *permanent* quarter section corner at the equidistant point on the *true* line, in the manner directed on the line between sections 25 and 26.

6. In this manner you will proceed with the survey of each successive section in the first tier, until you arrive at the north boundary of the township, which you will reach in running up a random line between sections 1 and 2. If this random line should not intersect at the corner established for sections 1, 2, 35 and 36, upon the township line, you will note the distance that you fall east or west of the same, from which distance you will calculate a course that will run a true line south to the corner from which your random started. Where the closing corner is on the base or standard line, a deviation from the general rule is explained under the head of "Diagram B."

7. The first tier of sections being thus laid out and surveyed, you will return to the south boundary of the township, and from the corner of sections 34 and 35 commence and survey the second tier of sections in the same manner that you pursued in the survey of the first, closing at the section corners on the first tier.

[24]

8. In like manner proceed with the survey of each successive tier of sections, until you arrive at the fifth tier; and from each section corner which you establish upon this tier, you are to run random lines to the corresponding corners established upon the range line forming the western boundary of

the township; setting, as you proceed, each *temporary* quarter section post at forty chains from the interior section corner, so as to throw the excess or deficiency of measurement on the extreme tier of quarter sections contiguous to the township boundary; and, on returning, establish the *true* line, and establish thereon the *permanent* quarter section corner.

QUARTER SECTION CORNERS, both upon north and south and upon east and west lines, are to be established at a point *equidistant* from the corresponding section corners, *except* upon the lines closing on the north and west boundaries of the township, and in those situations the quarter section corners will always be established at precisely *forty chains* to the north or west (as the case may be) of the respective section corners from which those lines respectively *start*, by which procedure the excess or deficiency in the measurements will be thrown, according to law, on the extreme tier of quarter sections.

Every north and south section line, except those terminating in the north boundary of the township, is to be eighty chains in length. The east and west section lines, except those terminating on the west boundary of the township, are to be within one hundred links of eighty chains in length; and the north and south boundaries of any one section, except in the extreme western tier, are to be within one hundred links of equal length. The meanders within each fractional section, or between any two meander posts, or of a pond or island in the interior of a section, must close within one chain and fifty links.

DIAGRAM A illustrates the mode of laying off township exteriors *north* of the BASE line and EAST and WEST of the principal MERIDIAN, whether between the base and first standard, or between any two standards; and the same general principles will equally apply to townships *south* of the base line and east and west of the meridian, and between any two standards *south*, where the distances between the base and first standard, and between the standards themselves, are five townships or thirty miles.

[25]

DIAGRAM B indicates the mode of laying off a TOWNSHIP into sections and quarter sections, and the accompanying set of field notes (marked B) critically illustrate the mode and order of conducting the survery under every variety of circumstance shown by the topography on the diagram. In townships lying south of and *contiguous* to the base or to any standard parallel, the lines between the northern tier of sections will be run *north*, and be made to close as *true* lines; quarter section corners will be set at forty chains, and section corners established at the intersection of such section lines with the base or standard, (as the case may be,) and the distance is to be measured and entered in the field book to the nearest corner on such standard or base.

DIAGRAM C illustrates the mode of making mound, stake, or stone corner boundaries for townships, sections, and quarter sections.

The mode and order of surveying the *exterior* boundaries of a township are illustrated by the specimen field notes marked A; and the mode and order of *subdividing* a township into sections and quarter sections are illustrated by the specimen field notes marked B. The attention of the deputy is particularly directed to these specimens, as indicating not only the method in which his work is to be conducted, buy also the order, manner, language, &c., in which his field notes are required to be returned to the Surveyor General's office; and such specimens are to be deemed part of these instructions, and any *departure* from their details, without special authority, in cases where the circumstances are analogous in practice, *will be regarded as a violation of his contract and oath*.

The subdivisions of fractional sections into forty acre lots, (as near as may be,) are to be so laid down on the official township plat in *red* lines, as to admit of giving to each a specific designation, if possible, according to its relative position in the fractional section, as per examples afforded by diagram B, as well as by a number, in all cases where the lot cannot properly be designated as a quarter-quarter. Those fractional subdivision lots which are not susceptible of being described according to relative local position, are to be numbered in regular series; No. 1 being (wherever practicable, and as a general rule) either the northeastern or the most easterly fractional lot, and proceeding from east to west and from west to east, alternately, to the end of the series; but such general rule is departed from under circumstances given as examples in fractional sections 4, 7, 19 and 30, where No. 1 is the interior lot of the northern and western tiers of the quarter sections to which there is a corresponding No. 2 given to the exterior lot, and the series of num-

[26]

bers is in continuation of the latter. The lots in the extreme northern and western tiers of quarter sections, containing either more or less than the regular quantity, are always to be numbered as per example. Interior lots in such extreme tiers are to be *twenty* chains wide, and the excess or deficiency of measurement is always to be thrown on the exterior lots; elsewhere, the assumed subdivisional corner will always be a point equidistant from the established corners.

The official township plat to be returned to the General Land Office is to show on its face, on the right hand margin, the meanders of navigable streams, islands, and lakes. Such details are wanted in the adjustment of the surveying accounts, but may be omitted in the copy of the township plat to be furnished to the district land office by the surveyor general. A suitable margin for *binding* is to be preserved on the left hand side of each plat. Each plat is to be certified, with table annexed, according to the forms subjoined to "diagram B," and is to show the areas of public land, of private surveys, and of water, with the aggregate area as shown on the diagram.

Each township plat is to be prepared in *triplicate*: one for the General Land Office, one for the district office, and the third to be retained as the record in the office of the Surveyor General.

The original field books, each bearing the *written approval* of the Surveyor General, are to be substantially bound into volumes of suitable size, and retained in the surveyor general's office, and certified *transcripts* of such field books (to be of *foolscap* size) are to be prepared and forwarded, from time to time, to the General Land Office.

With the copy of each township plat furnished to a district land office, the surveyor general is required by law to furnish *descriptive notes* as to the character and quality of the soil and timber found on and in the vicinity of each surveyed line, and giving a description of each corner boundary.

Printed blank forms for such notes will be furnished by the General Land Office. The forms provide eighteen spaces for *meander corners*, which, in most cases, will be sufficient; but when the number shall exceed eighteen, the residue will have to be inserted on the face of the township plat, to be furnished to the register of the district land office. There is shown a series of meander corners on diagram B, viz: from No. 1 to No. 22, on the river and islands; 23 to 28 being on Island lake; 29 and 30 on Clear lake; and 31 and 32 on lake in section 26.

There is also a distinct series of numbers, 1 to 7, to designate corners D. Reed's private survey, and to fractional sections, made such thereby; and the same series is continued from 8 to 14 inclusive, to

[27 [

designate corners to S. William's private survey, and to fractional sections made such thereby. These are numberings on the plat merely for the purpose of ready reference to the descriptions of such corners to be furnished to the registers.

The *letters* on "diagram B," at the "corners" on the township boundaries, are referred to in the descriptive notes to be furnished to the district land office, but are not required to be inserted on the official plat to be returned to the General Land Office.

The following chapter, on the subject of the variation of the magnetic needle, is extracted from the revised edition of the work on surveying by CHARLES DAVIES, L. L. D., a graduate of the Military Academy at West Point. The work itself will be a valuable acquisition to the deputy surveyor; and his attention is particularly invited to the following chapter, which sets forth the modes by which the variation may be ascertained.

[28]

VARIATION OF THE NEEDLE.

1. The angle which the magnetic meridian makes with the true meridian, at any place on the surface of the earth, is called the *variation of the needle* at that place, and is east or west, according as the north end of the needle lies on the east or west side of the true meridian.

2. The variation is different at different places, and even at the same place it does not remain constant for any length of time. The variation is ascertained by comparing the magnetic with the true meridian.

3. If we suppose a line to be traced through those points on the surface of the earth, where the needle points directly

north, such a line is called the *line of no variation*. At all places lying on the east of this line, the variation of the needle is west; at all places lying on the west of it, the variation is east.

4. The public is much indebted to Professor Loomis for the valuable results of many observations and much scientific research on the dip ad variation of the needle, contained in the 39th and 42d volumes of Siliman's Journal.

The variation at each place was ascertained for the year 1840; and by a comparison of previous observations and the application of known formulas, the annual motion, or change in variation, at each place, was also ascertained, and both are contained in the tables which follow.

5. If the annual motion was correctly found, and continues uniform, the variation at any subsequent period can be ascertained by simply multiplying the annual motion by the number of years, and adding the product, in the algebraic sense, to the variation in 1840. It will be observed that all variations west are designated by the plus sign; and all variations east, by the minus sign. The annual motions being all west, have all the plus sign.

6. Our first object will be to mark the line, as it was in 1840, of *no variation*. For this purpose we shall make a table of places lying near this line.

PLACES NEAR THE LINE OF NO VARIATION.

Place.	Latitude.	Longitude.	Variation.	An. Motion.
A Point..............	40°53′	80°13′	0°00′	+4′.4
Cleveland, Ohio......	41 31	81 45	− 0 19	4 .4
Detroit, Mich	42 24	82 58	−1 56	4 .
Mackinaw	45 51	84 41	−2 08	3 .9
Marietta, Ohio.......	39 0	81 28	−1 24	4 .3
Charlottesville, Va ...	39 02	78 30	+ 0 19	3 .7
Charleston, S. C......	32 42	80 04	−2 44	1 .3

[29]

At the point whose latitude is 40° 53′, longitude 80° 13′, the variation of the needle was nothing in the year 1840, and the direction of the line of no variation, traced north, was N. 24° 35′ west. The line of no variation, prolonged, passed a little to the east at Cleveland, in Ohio—the variation there being 19 minutes east. Detroit lay still further to the west of this line, the variation there being 1° 56′ east; and Mackinaw still further to the west, as the variation at that place was 2° 08′ east.

The course of the line of no variation, prolonged southerly, was S. 24° 35′ E. Marietta, Ohio, was west of this line—the variation there being 1° 24′ east. Charlottesville, in Virginia, was a little to the east of it—the variation there being 19′ west; whilst Charleston, in South Carolina, was on the west—the variation there being 2° 44′ east.

From these results, it will be easy to see about where the line of no variation istraced in our own country.

7. We shall give two additional tables:

PLACES WHERE THE VARIATION WAS WEST.

Places.	Latitude.	Longitude.	Variation.	An. Motion.
Angle of Maine	48°00′	67°37′	+19°30′	+8°.8
Waterville, Me.	44 27	69 32	12 36	5 .7
Montreal	45 31	73 35	10 18	5 .7
Keesville, N. Y.......	44 28	73 32	8 51	5 .3
Burlington, Vt........	44 27	73 10	9 27	5 .3
Hanover, N. H........	43 42	72 14	9 20	5 .2
Cambridge, Mass.	42 22	71 08	9 12	5
Hartford, Ct..........	41 46	72 41	6 58	5
Newport, R. I.........	41 28	71 21	7 45	5
Geneva, N. Y.........	42 52	77 03	4 18	4 .1
West Point...........	41 25	74 00	6 52	4
New York City.......	40 43	71 01	5 34	3 .6
Philadelphia	39 57	75 11	4 08	3 .2
Buffalo, N. Y.........	42 52	79 06	1 37	4 .1

PLACES WHERE THE VARIATION WAS EAST.

Places.	Latitude.	Longitude.	Variation.	An. Motion.
Mouth of Columbia River ...	46°12′	123°30′	−21°40′	Unknown.
Jacksonville, Ill.	39 43	90 20	8 28	+2′ .5
St. Louis, Mo..	38 37	90 17	8 37	2 .3
Nashville, Tenn.	36 10	86 52	6 42	2
Louisiana, at...............	29 40	94 00	8 41	1 .4
Mobile, Ala.	30 42	88 16	7 05	1 .4
Tuscaloosa, Ala.............	33 12	87 43	7 26	1 .6
Columbus, Geo.	32 28	85 11	5 28	2
Milledgeville, Geo.	33 07	83 24	5 07	2 .4
Savannah, Geo.	32 05	81 12	4 13	2 .7
Tallahassee, Fla............	30 26	84 27	5 03	1 .8
Pensacola, Fla..............	30 24	87 23	5 53	1 .4
Logansport, Ind.............	40 45	86 22	5 24	2 .7
Cincinnati, Ohio	39 06	84 27	4 46	3 .1

[30]

METHODS OF ASCERTAINING THE VARIATION.

8. The best practical method of determining the true meridian of a place, is by observing the north star. If this star were precisely at the point in which the axis of the earth, prolonged, pierces the heavens, then, the intersection of the vertical plane passing through it and the place, with the surface of the earth, would be the true meridian. But the star being at a distance from the pole, equal to 1° 30′ nearly, it performs a revolution about the pole in a circle, the polar distance which is 1° 30′: the time of revolution is 23 h. and 56 min.

To the eye of an observer, this star is continually in motion, and is due north but twice in 23 h. 56 min; and is then said to be on the meridian. Now, when it departs from the meridian, it apparently moves east or west, for 5 h. and 59 m., and then returns to the meridian again. When at its greatest distance from the meridian, east or west, it is said to be at its greatest *eastern* or *western* elongation.

The following tables show the times of its greatest eastern and western elongations.

EASTERN ELONGATIONS.

Days.	April.	May.	June.	July.	August.	Sept.
	H. M.	H. M.	H. M.	H. M.	H. M.	H. M.
1	18 18	16 26	14 24	12 20	10 16	8 20
7	17 56	16 03	14 00	11 55	9 53	7 58
13	17 34	15 40	13 35	11 31	9 30	7 36
19	17 12	15 17	13 10	11 07	9 08	7 15
25	16 49	14 53	12 45	10 43	8 45	6 53

WESTERN ELONGATIONS.

Days.	Oct.	Nov.	Dec.	Jan.	Feb.	March.
	H. M.	H. M.	H. M.	H. M.	H. M.	H. M.
1	18 18	16 22	14 19	12 02	9 50	8 01
7	17 56	15 59	13 53	11 36	9 26	7 38
13	17 34	15 35	13 27	11 10	9 02	7 16
19	17 12	15 10	13 00	10 44	8 39	6 54
25	16 49	14 45	12 34	10 18	8 16	6 33

[31]

The eastern elongations are put down from the first of April to the first of October; and the western, from the first of October to the first of April; the time is computed from 12 at noon. The western elongations in the first case, and the eastern in the second, occurring in the daytime, cannot be used. Some of those put down are also invisible, occurring in the evening, before it is dark, or after daylight in the morning. In such case, if it be necessary to determine the meridian at that particular season of the year, let 5 h. and 59 m. be added to, or subtracted from, the time of greatest eastern or western elongation, and the observation be made at night, when the star is on the meridian.

9. The following table exhibits the angle which the meridian plane makes with the vertical plane passing through the pole-star, when at its greatest eastern or western elongation: such angle is called the *azimuth*. The mean angle only is put down, being calculated for the first of July of each year:

AZIMUTH TABLE.

Year.	Lat. 32° Azimuth.	Lat. 34° Azimuth.	Lat. 36° Azimuth.	Lat. 38° Azimuth.	Lat. 40° Azimuth.	Lat. 42° Azimuth.	Lat. 44° Azimuth.
1851	1° 45½′	1° 48′	1° 50½′	1° 53½′	1° 56¾′	2° 00¼′	2° 04¼′
1852	1° 45′	1° 47½′	1° 50′	1° 53′	1° 56¼′	1° 59¾′	2° 03¾′
1853	1° 44½′	1° 47′	1° 49¾′	1° 52½′	1° 55¾′	1° 59¼′	2° 03¾′
1854	1° 44¼′	1° 46½′	1° 49¼′	1° 52′	1° 55¼′	1° 59′	2° 02¾′
1855	1° 43¾′	1° 46¼′	1° 48¾′	1° 51¾′	1° 54¾′	1° 58½′	2° 02¼′
1856	1° 43¼′	1° 45¾′	1° 48¼′	1° 51¼′	1° 54½′	1° 58′	2° 01¾′
1857	1° 43′	1° 45¼′	1° 48′	1° 50¾′	1° 54′	1° 57½′	2° 01¼′
1858	1° 42½′	1° 44¾′	1° 47½′	1° 50¼′	1° 53½′	1° 57′	2° 00¾′
1859	1° 42′	1° 44½′	1° 47′	1° 49¾′	1° 53′	1° 56½′	2° 00¼′
1860	1° 41¾′	1° 44′	1° 46½′	1° 49½′	1° 52½′	1° 56′	2° 00′
1861	1° 41¼′	1° 43¾′	1° 46¼′	1° 49′	1° 52¼′	1° 55¾′	1° 59½′

The use of the above tables, in finding the true meridian, will soon appear.

[32]

TO FIND THE TRUE MERIDIAN WITH THE THEODOLITE.

10. Take a board, of about one foot square, paste white paper upon it, and perforate it through the centre: the diameter of the hole being somewhat larger than the diameter of the telescope of the theodolite. Let this board be so fixed to a vertical staff as to slide up and down freely; and let a small piece of board, about three inches square, be nailed to the lower edge of it, for the purpose of holding a candle.

About twenty-five minutes before the time of the greatest eastern or western elongation of the pole-star, as shown by the tables of elongations, let the theodolite be placed at a convenient point and levelled. Let the board be placed about one foot in front of the theodolite, a lamp or candle placed on the shelf at its lower edge; and let the board be slipped up or down, until the pole-star can be seen through the hole. The light reflected from the paper will show the cross hairs in the telescope of the theodolite.

Then, let the vertical spider's line be brought exactly upon the pole-star, and, if it is an eastern elongation that is to be observed, and the star has not yet reached the most easterly point, it will move from the line towards the east, and the reverse when the elongation is west.

At the time the star attains its greatest elongation, it will appear to coincide with the vertical spider's line for some time, and then leave it, in the direction contrary to its former motion.

As the star moves towards the point of greatest elongation, the telescope must be continually directed to it, by means of the tangent-screw of the vernier plate; and when the star has attained its greatest elongation, great care should be taken that the instrument be not afterwards moved.

Now, if it be not convenient to leave the instrument in its place until daylight, let a staff, with a candle or small lamp upon its upper extremity, be arranged at thirty or forty yards from the theodolite, and in the same vertical plane with the axis of the telescope. This is easily effected, by revolving the vertical hair about its horizontal axis without moving the vernier plate, and aligning the staff to coincide with the vertical hair. Then mark the point directly under the theodolite; the line passing through this point and the staff, makes an angle with the true meridian equal to the azimuth of the pole-star.

From the table of azimuths, take the azimuth corresponding to the year and nearest latitude. If the observed elongation was east, the true meridian lies on the west of the line which has been found, and makes

[33]

with it an angle equal to the azimuth. If the elongation was west, the true meridian lies on the east of the line; and, in either case, laying off the azimuth angle with the theodolite, gives the true meridian.

TO FIND THE TRUE MERIDIAN WITH THE COMPASS.

11. 1. Drive two posts firmly into the ground, in a line nearly east and west; the uppermost ends, after the posts are driven, being about three feet above the surface, and the posts about four feet apart: then lay a plank, or piece of timber three or four inches in width, and smooth on the upper side, upon the posts, and let it be pinned or nailed, to hold it firmly.

2. Prepare a piece of board four or five inches square, and smooth on the under side. Let one of the compass-sights be placed at right angles to the upper surface of the board, and let a nail be driven through the board, so that it can be tacked to the timber resting on the posts.

3. At above twelve feet from the stakes, and in the direction of the pole-star, let a plumb be suspended from the top of an inclined stake or pole. The top of the pole should be of such a height that the pole star will appear about six inches below it; and the plumb should be swung in a vessel of water to prevent it from vibrating.

This being done, about twenty minutes before the time of elongation, place the board, to which the compass sight is fastened, on the horizontal plank, and slide it east or west, until the aperture of the compass-sight, the plumb-line, and the star, are brought into the same range. Then if the star depart from the plumb-line, move the compass-sight east or west along the timber, as the case may be, until the star shall attain its greatest elongation, when it will continue behind the plumb-line for several minutes, and will then recede from it in the direction contrary to its motion before it became stationary. Let the compass-sight be now fastened to the horizontal plank. During this observation it will be necessary to have the plumb-line lighted: this may be done by an assistant holding a candle near it.

Let now a staff, with a candle or lamp upon it, be placed at a distance of thirty or forty yards from the plumb-line, and in the same direction with it and the compass-sight. The line so determined makes, with the true meridian, an angle equal to the azimuth of the pole-star; and from this line the variation of the needle is readily determined, even without tracing the true meridian on the ground.

Place the compass upon this line, turn the sights in the direction of it, and note the angle shown by the needle. Now, if the elongation, at

[34]

the time of observation, was west, and the north end of the needle is on the west side of the line, the azimuth, plus the angle shown by the needle, is the true variation. But should the north end of the needle be found on the east side of the line, the elongation being west, the difference between the azimuth and the angle would show the variation, and the reverse when the elongation is east.

1. Elongation west, azimuth---------------- 2° 04'
   North end of the needle on the west,
   angle -- 4° 06'

        Variation ------------------------ 6° 10' west.

2. Elongation west, azimuth---------------- 1° 59'
   North end of the needle on the east,
   angle -- 4° 50'

        Variation .... 2° 51' east.

3. Elongation east, azimuth ---------------- 2° 05'
   North end of the needle on the west,
   angle -- 8° 30'

        Variation ------------------------ 6° 25' west.

4. Elongation east, azimuth ---------------- 1° 57'
   North end of the needle on the east,
   angle -- 8° 40'

        Variation---------------------- 10° 37' east.

REMARK I. The variation at West Point, in September, 1835, was 6° 32' west.

REMARK II. The variation of the needle should always be noted on every survey made with the compass, and then if the land be surveyed at a future time, the old lines can always be re-run.

12. It has been found by observation, that heat and cold sensibly affect the magnetic needle, and that the same needle will, at the same place, indicate different lines at different hours of the day.

If the magnetic meridian be observed early in the morning, and again at different hours of the day, it will be found that the needle will continue to recede from the meridian as the day advances, until about the time of the highest temperature, when it will begin to return, and at evening will make the same line as in the morning. This change is called the *diurnal variation*, and varies, during the summer season, from one-fourth to one-fifth of a degree.

[35]

13. A very near approximation to a true meridian, and consequently to the variation, may be had, by remembering that the pole-star very nearly reaches the true meridian, when it is in the same vertical plane with the star Alioth in the tail of the Great Bear, which lies nearest the four stars forming the quadrilateral.

The vertical position can be ascertained by means of a plumb-line. To see the spider's lines in the field of the telescope at the same time with the star, a faint light should be placed near the object-glass. When the plumb-line, the star Alioth, and the north star, fall on the vertical spider's line, the horizontal limb is firmly clamped, and the telescope brought down to the horizon; a light, seen through a small aperture in a board, and held at some distance by an assistant, is then moved according to signals, until it is covered by the intersection of the spider's lines. A picket driven into the ground, under the light, serves to mark the meridian line for reference by day, when the angle formed by it and the magnetic meridian may be measured.

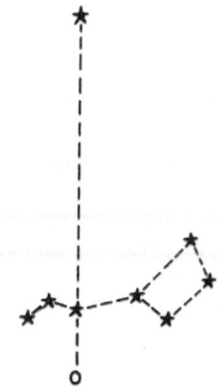

INDEX.

Referring the lines to the pages of the field-notes.

Town. 25 N. Range 2 W. Willamette Meridian.

A.

Field Notes of the survey of the exterior boundaries of Township 25 north of range 2 west of the Willamette meridian, in the Territory of Oregon, by Robert Acres, deputy surveyor, under his contract No. 1, bearing date the 2d day of January, 1854.

South Boundary, T. 25 N. R. 2 W. Willamette Meridian.

CHAINS.	
	Begin at the post, the established corner to Townships 24 and 25 North, in Ranges 2 and 3 West. The witness trees all standing, and agree with the description furnished me by the office, viz:
	A Black Oak, 20 in. dia. N. 37 E. 27 links,
	A Bur Oak, 24 in. dia. N. 43 W. 35 links,
	A Maple, 18 in. dia. S. 27 W. 39 links,
	A White Oak, 15 in. dia. S. 47 E. 41 links.
	East, on a random line on the South Boundaries of sections 31, 32, 33, 34, 35, and 36.
	Variation by Burt's improved solar compass, 18° 41' E,
	I set temporary half mile and mile posts at every 40 and 80 chains, and at 5 miles, 74 chains, 53 links, to a point 2 chains and 20 links north of the corner to Townships 24 and 25 North, Ranges 1 and 2 W.
	(Therefore the correction will be 5 chains, 47 links West, and 37 links *South* per mile,)
	I find the corner post standing and the witness trees to agree with the description furnished me by the surveyor general's office, viz:
	A Bur Oak, 17 in. dia. bears N. 44 E. 31 links,
	A White Oak, 16 in. dia. N. 26 W. 21 links,
	A Lynn, 20 in. dia. S. 42 W. 15 links.

Selected Bibliography

Brown, Curtis M., 1969, *Boundary control and legal principles*, John Wiley & Sons, New York.

Brown, Curtis M. and Eldridge, Winfield H., 1962, *Evidence and procedure for boundary location*, John Wiley & Sons, New York.

Brown, Lloyd A., 1949, *The story of maps*, Bonanza Books, New York.

Clawson, Marion, 1968, *The land system of the United States—an introduction to the history and practice of land use and land tenure*, Beard Books.

Holbrook, Stewart H., 1947, *The story of American railroads*, Crown Publishers, New York.

Madson, T.S. II and Seemann, Louis N.A., 1980, Fading footsteps, Land Surveyor's Seminar, University Station, Gainesville, FL 32604.

Manual of surveying instructions, 1855, 1871, 1881, 1890, 1894, 1902, 1930, 1947, and 1973; new edition planned 2009, U.S. Department of the Interior, Bureau of Land Management, U.S. Government Printing Office, Washington D.C.

Minnick, Roy, 1980, *A collection of original instructions to surveyors of the public lands*, Hallmark Enterprises, Rancho Cordova, CA 95670.

Moffit, Frances H. and Bossler, John D., 1998, *Surveying*, 10th Ed. Addison-Wesley Longman, New York.

Robillard, Walter G. and Bouman, Lane J., 1987, *Law of surveying and boundaries*, 5th Ed. The Michie Co., Charlotteville, VA.

Steward, Lowell O., 1935; *Public land surveys—history, instructions, methods*, The Meyer Printing Co., Minneapolis, MN 55415.

Tillotson, Ira M., 1973, *Legal principles of property boundary location on the ground in the public land survey states*, Ira M. Tillotson, Missoula, MT 59801.

Wheeler, Keith, 1973, *The railroaders*, Time-Life Books, New York.

White, C. Albert, 1983, *A history of the rectangular survey system*, U.S. Department of the Interior, Bureau of Land Management, U.S. Government Printing Office, Washington D.C. Stock Number 024-011-00150-6.

Surveying Roots

Surveying relies on many roots, from the legal principles of property ownership to mathematical principles derived from astronomy, to "Arabic" numbers to show its values.

The word *cadaster*, as used in the terms "cadastral law" or "cadastral surveyor," is expanded to reflect the generic definition of "pertaining to landed property as to its extent, value, and ownership" (*Merriam-Webster New Collegiate Dictionary*). Following this definition, the majority of the population, from property owner to real estate agent to (cadastral) lawyer to surveyor and, even by the strictest definition, the county assessor, should be knowledgeable about the material presented.

Cadastral surveys reflect the character of their time and are influenced by their heritage, the environment of their applications, and the constraints of the surveying equipment. This is especially true in the United States where people from many parts of the world contributed to the legal contexts of the law, the vastness of the land, which dictated many surveying methods (GLO surveys, transit and chain, etc.), and the inevitable progression of technology with its improvements of accuracies over the past. Perhaps strange to an outsider is the fact that cadastral surveys and laws are unique to each state and even though there are federal lands in each state, there are no federal "overall" cadaster laws.

The roots of the U.S. cadastral laws can be found in several places, and probably some of the oldest records are from Mesopotamia (now Iraq) where property transactions were written in cuneiform on clay tablets for some 2600 years at the time of King Nebuchadnezzar. The ancient Egyptians (starting about 3,000 BC) wrote extensive property records and surveying methods on papyrus, especially since there was a great need for resurveys after the annual flooding with water and mud by the Nile River.

The illustration below shows the Egyptian rope stretchers (i.e., surveyors) ready for work with a rope and knots tied at 3, 4, and 5 cubit intervals (see Figure 41).

How those ancient laws found their way into Roman law is not clear, but in essence Roman law is a very strong ancestor of French and Spanish law and also is part of English common law.

In order to work in our modern society, a person has to understand the principles of property laws and surveys based on the following:

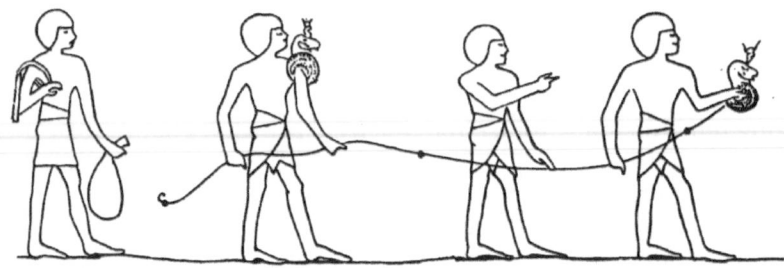

FIGURE 41
Egyptian rope stretchers.

English Common Law

Applicable in most of the United States, probably superimposed on some previous surveys, and comprised of:

Metes and bounds

Government Land Office Surveys (U.S. Rectangular Survey)

Mining laws

Torrens title system

French Law (Also Called the Napoleonic Code)

The Napoleonic code, derived from Roman law, is an integral part of the judicial system in the U.S. Gulf Coast states from Florida to Texas. Some of the original surveys along the Gulf Coast, primarily in Louisiana and on both the Canadian and U.S. side of the St. Lawrence River, New Hampshire, Maine, upstate New York, Ohio, Michigan, and some stretches along the Mississippi River, are French based. Most of the original French surveys were made before the Louisiana purchase in 1803. Valid grants and possessions were excluded from the public domain and thus remained in private ownership; their deeds however are still valid today.

> *French Tracts*—also called "long lots," were probably designed to offer as many people as possible a share of the good land and a frontage onto a waterway or lake for ease of transportation.
>
> *The Federal Act of March 3, 1811, 2Stat.662* (repealed December 16, 1930, 46Stat.1029) was in deference to the French settlers who

opposed the rectangular surveys and who wanted to buy land by their traditional method of 5 arpents (14.50 ch = 957 ft) frontage, 40 arpents (116.25 ch = 7,672.5 ft) in depth, or 200 arpents (168.56 acres) in area. It applied only in the Orleans Territory, to lands "adjacent to any river, lake, creek, bayou, or watercourse." The surveyors were to run the side lot lines at right angles to the general course of the watercourse, keep the side lines as nearly parallel as possible, and vary their length to provide common corners and to avoid gaps or gores.

Arpent: Land grants of the French crown were usually described in arpents, which is a unit of area ≈ 0.85 acre. The determination of length depends on whether the original grantor was of French or English descent:

In Arkansas and Missouri: 1 arpent = 0.8507 acre, 1 square arpent = 192.50 ft;

In Louisiana, Mississippi, Alabama, and northwest Florida: 1 arpent = 0.84725 acre, 1 square arpent = 191.994 ft.

(*Source*: Definition of Surveying Terms, ASCE & ACSM, 1972)

Figure 43 is a GLO township plat (T9N, R10E, Louisiana Principal Meridian), dated 1829. For people familiar with the regular "square" pattern of the mid-continent prairie states, this plat is unusual because it shows the "French tracts" more or less perpendicular to the watercourse (Lake St. John), abutting the "square" GLO sections, and the numbering of tracts and sections up to number 59. In general the government surveys respected private ownership of written record, and it is not possible to tell from this plat how many French tracts were originally made by early settlers or added by government surveyors under the 1811 Act. Some tracts are labeled with their private owner's name; others are numbered only and are presumed to be government land.

Figure 44 is a copy of a modern map of the same area, located at about mile marker 375 of the Mississippi River.

It is remarkable that during the 147 years since the original GLO plat was drawn, neither the surveying lines, nor the name of Fletchers Lake or Lake St. John have changed. Buckner Bayou still drains into Fletchers Lake, but the northern end of the lake has become a swamp called "Little Fletchers Lake," Bayou Tensas has become Little Tensas Bayou, etc.

It is also unfortunate that the modern maps do not show which of the monuments were found at the time the maps were field checked. (See U.S.G.S. 7½ m maps: Spokane, Louisiana [1976] and Waterproof, Louisiana [1994]).

Figure 45 is an abstract of the general property lines traced from the "official" U.S.G.S. topographical maps. As compared with the original, it dramatizes the fact that property lines remain virtually unchanged over time.

Spanish Law

Spanish property and water laws had a profound influence on the southern tier of states from Florida to California and northward into Colorado, Utah, and Nevada. Starting in about 1500 and until the treaty of 1819 with the United States, the Spanish legacy from Florida westward to the French settlements of Louisiana was minimal and restricted to place names and very small property holdings. On the other hand, Spanish and later Mexican influences left a permanent mark from Texas westward to California. Politically, the Spanish presence ended with the Mexican independence in 1821. In turn, this was followed by the Mexican era, which ended politically with the treaty of Guadalupe–Hidalgo in 1848 and the Gadsden Purchase in 1853.

Some original surveys are still valid to this day, and in the Southwest it is possible to have four sources of title: (1) Spanish, (2) Mexican, (3) Texan, and (4) United States.

The Spanish and Mexican codes, derived from Roman law, are called "Recopilacion de las Leyes de Indias," and the "Ordenanza de Intedentes" of October 15, 1754 had a marked influence on the affected states, especially Texas property law.

With the treaty of Guadalupe-Hidalgo (1848) the United States agreed to honor and confirm the "Rancho" land grants and any rights of the original inhabitants (Indians) made previously by the Spanish and Mexican governments, or so it seemed, and many stories abound. In 1891 the U.S. Congress authorized settlement of the land-grant claims by the Court of Private Land Claims. In essence, the courts had to deal with real property and the conflict between the Mexican and English jurisprudence is as follows:

> The Rancho land grants were granted to a person whose property was marked with the verbal understanding that "everybody knows where the property is," or if anything was written down, the records were kept with the family or in some far away archive. The Americans on the other hand required that a property be described in writing and that the records be kept by a local court. In the final analysis, relatively few of the grants were granted a U.S. patent or remained in private ownership.

Figure 47 shows some land grants in the Southwest. First, it should be noted that Spanish-Mexican land grants were only granted in Texas, New Mexico, southern Colorado, southern Arizona, and California. Second, a detailed study of each state's grants should be made because many of the grants made were either adjudicated by U.S. courts in parts only or later completely rejected.

Colorado Land Grants

Since Colorado was still a territory at the time, Governor Manual Armijo of New Mexico made six land grants in "Colorado" to attract settlers (see Figure 47):

Year	Colorado Portion of Grant	Original Claim (in acres)	Final Grant (in acres)
1832	Tierra Amarillo	82,000[a]	82,000[a]
1833	Conejos	2,000,000[a]	—
1841	Maxwell	250,000[a]	250,000[a]
1843	Vigil and St. Vrain	4,300,000	97,390
1843	Nolan	300,000[a]	38,000
1843	Sangre de Cristo	846,000[a]	846,000[a]
1860	Luis Maria Baca	100,000	100,000

[a] Rough estimates.

Additionally, in 1860 the U.S. government deeded about 100,000 acres (about 156 sq. mi.) to the Baca family as Luis Maria Baca Grant No. 4.

Figure 48 is the modern equivalent (1985) of the south-central portion of Colorado. The Louis Maria Baca Grant No. 4 is still intact; the Sangre de Cristo Grant name is still carried on the maps, but the land by now is subdivided into many private land parcels. In September 2004 the Colorado Nature Conservancy completed the last of a complex set of real estate transactions for $31.28 million to purchase the Baca Grant No. 4 with the vast majority of the money spent on the water rights. Finally in November 2004, for an additional $3.4 million from the federal government, the land (and water rights) for 97,000 acres out of the original 100,000 acres were purchased and are now used to enlarge the "new" Great Sand Dunes National Park and create the Baca National Wildlife Refuge. Verbally, the author has been told that the final ground surveys and title searches since the 1860s have not yet been completed.

Colorado's county lines generally follow the cardinal directions (north–south or east–west) or the irregular lines of the watersheds along mountain ridges. In south-central Colorado a "diagonal line" starts just south of Little Bear Peak (see Figure 48, south-central Colorado 1985). One of the most prominent mountain features of the San Louis Valley stands out as a modern legacy of the original Mexican Land grants. This original grant line starts at an unnamed peak, about 1200 ft south of Little Bear Peak, not on Blanca Peak as some maps show, and heads S 43°20' W to the center of the Rio Grande del Norte at the most easterly point of the La Loma del Norte. As a surveyor, standing on top adjacent to Little Bear Peak, it is easy to philosophize about the changes which have taken place since the original signing in 1843 of the Sangre de Cristo Land Grant in the Governor's Palace in Santa Fe, (now) New Mexico. Since many of the old surveys were drawn on natural landmarks, it does not look obvious that the line between Alamosa and Costilla County was ever physically surveyed in its entirety. Until the present it was probably sufficient, judging by existing fence lines, to survey the line with chain and transit in the San Louis Valley from about 7205 ft at the river, to maybe the 8,000+ ft elevation. Future generations probably will have to survey the entire 20.9 mile line all the way to an unnamed peak to the endpoint south of Little Bear Peak at 13,873 ft. It is easy enough to access the line in the valley, but despite Jeeps and helicopters, a surveyor still has to climb the summit adjacent to Little Bear Peak on foot, in order to use his GPS receiver. The hike to the top probably would be worthwhile, at least in summer, because to the west the headwaters of the Rio Grande River in the San Juan's are still as majestic as ever. Some 56 miles to the north the Sangre de Cristo Mountains appear to align with the Collegiate Range on Poncho Pass and more than 60 miles to the south are Santa Fe and the Rio Grande valley of central New Mexico. Especially at dawn, the curvature of the earth can still be seen on the prairie of eastern Colorado, which at the time of the original grant (1843) was still round. But within a scant 20 years, Alexander Clark redefined this curvature into the Clarke 1866 ellipsoid, which was only to be redefined about another 120 years later into the ellipsoid of 1980 (WGS '84).

> **"The retracement of the original surveys is, perhaps, the land surveyor's greatest problem today; consequently, it is the primary rule that a retracement surveyor follow in the footsteps of the original surveys.—"** Bascom Giles, Texas Land Commissioner

The Beginning of the Rectangular U.S. Survey System

The lure of instant wealth made people walk across the continent to the goldfields of California, and the U.S. Rectangular Survey System made the American dream of owning private land in an easy and orderly fashion possible. By providing clear title, the system made the buying and selling of land easy, whether by squatter, settler, or speculator, or by a modern family wanting their own plot of private land.

Historically, it is fascinating to see the evolution of the surveying principles. When and where permitted, property surveys started as a slowly growing metes and bounds "quilt pattern" in the Colonial states. Following the Revolutionary

War and subsequent Independence (1776) an enormous land rush to the West started. The metes and bounds with its interdependency on adjacent surveys was unable to cope, and the Rectangular Survey System with its effective "coordinates"—as, for example, section, township, range, and principal point—provided a unique location. The fact that every survey needed to go no more than a half mile to have a starting point where the Government survey had been made, anywhere in the country, made settlement possible. Land was cheap and was sold as sections (1 sq. mi. or 640 acres at $2/acre) or in quarter sections. Now fast forward to the present. Land has appreciated in value and instead of being sold by the section, it now is literally sold by the square foot and every part of the property perimeter is described by a metes and bounds description, which in turn is "anchored" or tied to the old GLO survey.

The path from Thomas Jefferson's (1743–1826) initial concepts as chairman of the Public Lands Commission to the present has not always been straight and uncontested. Jefferson, a great advocate of the decimal (and metric) system, envisioned a square grid pattern covering the land, 10 geographic miles square, subdivided into fractions of 10 and measured in meters. During his stay in France as U.S. ambassador, the township was reduced to 6 statute miles (following a New England pattern used in Massachusetts, Connecticut, and Rhode Island) and subdivided into sections of 1 statute mile each; the tradition of the practical Gunter chain prevailed. The Gunter chain lends itself to an easy subdivision into halves and quarters (i.e., 80 chains = 1 statute mile, one half again = 40 chains, half again = 20 chains, etc.) and 80 × 80 chains = 640 acres; 40 × 40 chains = 160 acres, etc. In the late 1700s, it was easier to lay out, for example, a road easement of 1½ chains = 99 ft, rather than 100 ft with a "decimal chain," which had not yet been invented. When on May 20, 1785 Congress passed the Land Ordinance, its members had more important items than the practicality of the Gunter chain on their minds. Foremost was the need to raise money to pay for the Revolutionary War, since without the power to tax, land sales were the most obvious source of revenue, especially since most states had deeded to the new federal government vast amounts of western land as a condition of joining the union. Another pressing need was to honor the land bounties promised to Revolutionary War veterans, which precipitated a very long discussion on how to best survey the land. The Land Ordinance of 1785 addressed most of the concerns, plus two additional points, namely that the land had to be surveyed before it was opened for sale and that at least conceptually the land was to be sold to individuals rather than to large land companies; as later history would show, neither of those restrictions was always met.

Stories of small and large area land transactions, both legitimate and fraudulent, abound. It finally cumulated in the Georgia land scandal and land panic of 1796–97 and the indictment of a U.S. Supreme Court Justice (James Wilson). From the onset, the U.S. Congress was closely involved in the management of federal lands, and major land legislation was passed by Congressional acts in 1796, 1800, 1804, 1805, 1812, 1832, 1899, 1909, 1910, 1925, and 1926. Since then, control has shifted from legislative to regulative action through the BLM (Bureau of Land Management), which in 1946 moved from the Department of the Treasury to the Department of the Interior. The agency is responsible for managing federal lands, about 264 million acres (about 4,112,500 sq. mi.) and its natural resources.

Patterned after the U.S. Land Survey System, the Canadian rectangular system started on July 10, 1871, and is called the Canadian Dominion Survey (DLS). The most visible difference between the U.S. and Canadian surveys is the numbering of the sections, as shown in Figure 15 and Figure 16 in Chapter 2. In some townships, mostly surveyed from 1871 to 1880, each section is surrounded by a road allowance (i.e., a 1½ chain (30.2 m) "space" between sections for a road right-of-way), later reduced to 1 chain (20.12 m); instead of just 6 miles, each of these townships is either 6 miles + 9 chains or 6 miles + 6 chains on each side.

A Brief History of Some Surveys in Ohio

The lands of what is now the state of Ohio served as a testing ground for various surveying and land sale schemes. Ultimately, the state was subdivided into 19 different grants with many details surviving to the present day. The following examples, only to show the variety, are presented in an abstracted form:

The Seven Ranges ("The Old Seven Ranges") and Geographer's Line—Surveyed August 1784 to June 1787 (see Figure 50)

- The very first rectangular survey in the U.S., surveyed under the ordinance of 1785.

- Point of origin; defined as the intersection of the western boundary of Pennsylvania and the "northern" shoreline of the Ohio River. The position, identified by a wooden stake in the original survey has been lost; a commemorative monument set in 1960 is 1,112 ft (more or less) north of the original stake. For an illustration see page ii, C. Albert White's *A History of the Rectangular Survey System,* and page 193, *Initial Points of the Rectangular Survey System* by the same author. This was the first point of origin surveyed in the U.S. There is no special significance attached to it, otherwise. It was followed by 37 other principal points in the next 165-some years.

- The geographer's line, later called "baseline," was surveyed under the supervision of Thomas Hutchins, the first geographer of the United States. The line was run westward with a sextant, common (magnetic) compass, a circumferentor (to measure angles), and a two pole (33 ft) Gunter chain held horizontally. Even though required by the 1785 ordinance to be run as a true geographic line, the line was run as a line with a magnetic bearing and the compass was not corrected for declination.

- By June 1797, the geographer's line was finally extended for its full 42 mile length, i.e., 7 × 6 miles/township = 42 miles. Section 16 was dedicated to schools and Sections 8, 11, 26, and 29 were reserved for the federal government.
- Guide meridians were run south from the geographer's line with full 6 mile townships until the Ohio River was intersected, leaving a fractional township; then starting at the fractional township with 1, the rest of the townships were numbered northward from the Ohio River back to the geographer's line.
- Land sold at public auction at $1.60/acre.

Ohio Company Purchase—Surveyed April 1788 to 1796 (?) (see Figure 50 and Figure 51)

- No point of origin.
- Company organized by General Rufus Putnam and others in 1786–87. When General Putnam later became Surveyor General of the United States, he initiated the contract system for surveyors which set a fixed price per mile surveyed (initially $2 or $3 per mile!). The contract system was discontinued about 1910.
- Extended the surveys westward and southward from "The Old Seven Ranges" and followed the pattern of numbering townships, ranges, and section numberings:
- Section 16 set aside for schools
- Section 29 set aside for religious organizations
- Township 8 & 9, Range 14, reserved for a college (about 72 sq. mi.!) and since 1804 the main campus of Ohio University in Athens, Ohio
- The federal government reserved for each township a third of all the gold, silver, copper, and lead deposits found. In addition, all salt springs and salt deposits were reserved.
- 1,500,000 acre tract (about 2343.75 sq. mi.), land ultimately was sold at 12¢ cents/acre

The Donation Tract—Surveyed 1792 (?) (see Figure 50 and Figure 51)

- When the Ohio Company encountered financial problems it requested that Congress donate 100,000 acres to the company in trust. The company was supposed to donate 100 acres to any male who would settle in the tract.
- Followed "The Old Seven Ranges" pattern of numbering townships, ranges, and sections, but most of the land was never subdivided into sections

The Symmes Purchase—Surveyed Late 1788 to (?) (see Figure 51)

- Named after Ceves Symmes, who contributed much of his personal fortune in support of the Continental Army and who was looking for some compensation. The purchase was so badly managed and the surveys were so badly executed by private surveyors, that it clearly showed the necessity to have the land surveyed by government surveyors before it was put on the market for sale.
- About 1,000,000 acres (about 1562.5 sq. mi.).
- Established a "baseline and meridian line."

    Established "Fractional Range 1" and "Fractional Range 2," and Ranges 1, 2, and 3. Ranges later extended northward from Range 4 to 9 by survey "Between the Miamis" (i.e., the land between the Great Miami and the Little Miami rivers).

    Townships numbered eastward from Great Miami River to Little Miami River.

- The only survey in the U.S. where ranges are numbered north–south and townships east–west.

The Virginia Military Tract—Surveyed August 1790 until Mid-1800s (see Figure 51)

- Land set aside north of the Ohio River and between the Little Miami and Scioto Rivers. Area surveyed basically by metes and bounds surveys. Claimants paid for cost of private surveys.
- Granted land to soldiers from Virginia.

    100 acres for soldier or sailor with at least 3 years' service

    15,000 acres to a major general.

- Original grant 4 million acres (about 6250 sq. mi.); ultimately only about 76,735 acres (about 119.8 sq.mi.) claimed. The remaining land reverted to State of Ohio.

The Connecticut Western Reserve—Surveyed 1796 to (?) (see Figure 51)

- Unlike most other colonial states when it joined the union, the Connecticut Western Reserve encompassed about 3,667,000 acres (about 5729.69 sq. mi.) from the west boundary of Pennsylvania 120 miles westward and from the 41°N parallel north to the shores of Lake Erie.

- Baseline = 41°N parallel.

    Ranges numbered from 1 to 24 westward from Pennsylvania state line.

    Townships numbered northward from baseline to the shores of Lake Erie.

    Townships are a "5 × 5 mile square."

    Starting in the NE corner, townships are subdivided either into 25, 1 × 1 mile sections (640 acres each) or 50 lots (320 acres each).

The Firelands—Surveyed 1796 to ? (see Figure 51)

- 500,000 acres (about 781.25 sq. mi.) given by Connecticut to settlers after the British burned New Haven, Greenwich, Norwalk, Fairfield, and New London, Connecticut.

- Range and township designation is a continuation of the Connecticut Western Reserve pattern.

- 5 × 5 mile Townships are subdivided into quarter townships (about 4,000 acres) and numbered with SE 1/4 = 1, NE 1/4 = 2, etc., and then into lots.

Historical Notes—A Summary

The historical land acquisitions and partial disposition of public lands by the United States require some comments (see Figure 49 [source: U.S. Bureau of Land Management], Table 3.1, and Table 3.2).

First, it must be realized that both the map and the tables, even though they appear to be accurate, represent generalizations only. During the time span from about 1700 to about 1860 land had little value, and there were no physical ground surveys which delineated most of the purchases or treaties. It was not until the late 1920s that some state boundaries were finally legalized.

Starting with the six colonial Eastern seaboard states (Virginia, North Carolina, South Carolina, Connecticut, Massachusetts, and Georgia) their original charters had provisions for western lands "west to the South Sea." As part of admission to the Union, those western lands were ceded to the United States between 1780 and about 1795 and later became all or parts of Wisconsin, Michigan, Illinois, Indiana, Ohio, Mississippi, and Alabama. The Carolina land grant was split into North and South Carolina in 1729, and in 1789 the present state of Tennessee was ceded to the United States. Likewise, the Virginia grants were reduced in size, and in 1792 Kentucky became a state. In 1866 West Virginia was separated. Not unlike the governments, land barons were also

TABLE 3.1

Land Acquisition by the United States

Name	From	Date	Area in sq. mi.	Area in acres	Cost in $ million	Cost in $/acre	Present states (or portions of states)
Louisiana Purchase	France	April 30, 1803	827,192	529	15	0.03	Arkansas, Colorado, North and South Dakota, Iowa, Louisiana, Mississippi, Missouri, Montana, Nebraska, Oklahoma, Wyoming
Florida	Spain	Feb. 22, 1819	72,101	46	6.7	0.14	Florida, western Louisiana
Texas	Texas enters union	Dec. 29, 1845	266,807	171			Texas
Oregon Territory	British, Russia	1819 to 1848	286,541	183			Washington, Oregon, Idaho, Wyoming, Montana
Treaty of Guatalupe	Mexico	Feb. 2, 1848	529,189	19	16.3	0.05	Arizona, New Mexico, California, Nevada, Utah, western Colorado
Texas Purchase	Texas	1850	123,281	78.9	15.5	0.20	New Mexico, eastern Colorado, Wyoming
Hawaii	France, Britain	1851	6,425	4.1	4	0.97	Hawaii
Gadsden Purchase	Mexico	1853	29,670	19.0	10	0.53	Arizona, New Mexico
Alaska	Russia	Mar. 30, 1867	586,400	375	7.2	0.02	Alaska
Midway Island	Occupied	Sep. 30, 1867	2	1280			U.S. protectorate
Ute Indian Purchase	Ute Indians	1868	1456	932,153			Land located in central Colorado
Philippine Islands	Spain	Dec. 10, 1898	115,831	74.1	20	0.27	Independence from U.S., July 4, 1946
Puerto Rico	Spain	1898	3,435	2.2			Self-governing
Guam	Spain	1898	212	135,680			Self-governing
Wake Island	Spain	Jan. 17, 1899	3	1,920			U.S. protectorate
American Samoa	Britain, Germany	Dec. 2, 1899	76	48,640			U.S. protectorate
Panama Canal Zone	Panama	Feb. 26, 1904	553	353,920	10	28.25	Independence from U.S. on Dec. 31, 1999
Virgin Islands	Denmark	Jan. 12, 1917	133	85,120	25	0.29	Republican-type government

TABLE 3.2

Disposition of Public Lands

Land Use Allocation	Acres in Millions	Sq. mi. (per 1,000)	%
Misc.[a]	303.5	474.2	13.1
Homesteads	287.5	449.2	12.4
Railroads	94.4	147.5	4.1
Veterans	61.0	95.3	2.6
Private claims[b]	34.0	53.1	1.5
Timber and stone[c]	13.9	21.7	0.6
Timber culture[d]	10.9	17.0	0.5
Desert land[e]	10.7	16.7	0.5
Total U.S.	**2,318.7**	**3,623**	
Total disposed	**815.9**	**1,274.8**	**35.2**
Allocations to States			
Schools	77.6	121.2	23.6
Swamps	64.9	101.4	19.8
RR construction	37.1	58	11.3
Institutions[f]	21.7	33.9	6.6
Misc.[g]	117.6	183.8	35.8
Canals and rivers	6.1	9.5	1.9
Wagon roads	3.4	5.3	1.0
Total disposed	**328.4**	**513.1**	

Source: Bureau of Land Management, 1994.

[a] Chiefly public, private and preemption sales, but includes mineral entries, scrip locations, sales of townships, and townlots.

[b] The government has confirmed title to land claimed under valid grants by foreign governments prior to the acquisition of the public domain.

[c] The law provides for the sale of lands valuable for timber or stone unfit for cultivation.

[d] The law provides for the granting of public lands to settlers on condition that they plant trees.

[e] The law provides for the sale of arid agricultural lands to settlers who irrigate them and bring them under cultivation.

[f] Universities, hospitals, asylums, etc.

[g] For the construction of various public improvements,

into land transactions; for example, William Penn bequested 27,000,000 acres (about 42,200 sq. mi.) to his sons. George Washington only bequested 49,000 acres (about 76.5 sq. mi.) to his heirs in the Kanawah River Valley of West Virginia. By a modern comparison, one of the largest cattle ranches in the country, the King Ranch in Texas, is "only" 825,000 acres (1,289 sq. mi.).

Following, in an abstract format and not necessarily in chronological order, are the major additions to U.S. territory.

As stated previously, the Louisiana Purchase in 1803 must unquestionably rate as one of the largest metes and bounds bargains of history. As described, the boundaries from the Gulf Coast north to the 49th parallel (still the present border with Canada) were well defined. "From the Mississippi-River west" was probably rather well defined below St. Louis, Missouri, but became progressively less defined northward past Minneapolis–St. Paul. But "westward to the continental divide" can only be excused in the context that neither the French nor the U.S. signatories had the foggiest idea where the continental divide was (is) located. Probably everybody agreed that it most probably was between St. Louis and San Francisco, but then again there might be the Northwest Passage around the northern waters of the continent.

The Lewis and Clark expedition (May 14, 1804 to September 23, 1806) established only a few points on the Continental Divide when they crossed over it in Idaho and Montana. At an original cost of 3¢/acre, it probably was not that critical that the land covered by the Louisiana Purchase covered the "old" country" equivalent of Norway, Sweden, Great Britain, France, Germany, the low countries, Italy and Switzerland, and Austria added for good measure!

By treaty with Great Britain, the "Basin of the Red River of the North" was acquired in 1818 and the Oregon territory was added by treaties with Spain (1819), Great Britain (1818 and 1846), and Russia (1824).

Land in the southwest (all or parts of California, Arizona, New Mexico, Texas, Utah, Colorado, Wyoming, Kansas, and Oklahoma) can have title from four sovereign sources: Spanish, Mexican, Texas, or the United States. The Spanish era began in the early 1500s and ended with the Mexican independence in 1821. The Mexican era in turn ended with the treaty of Guadalupe–Hidalgo in 1848 and the Gadsden Purchase in 1853. When Texas entered the Union in 1845, about 122,805 acres outside the present state were ceded to the United States, and the entire state of Texas remains a "metes and bounds state" only.

Part of Colorado fell under the Louisiana Purchase (1803), and another part was acquired from Texas in 1850 and from Mexico in 1853; the parcel in "central" Colorado was purchased from the Ute Indian tribe in 1868.

The Florida panhandle and western Florida as far west as Mobile, Alabama, as well as the land in western Louisiana, was ceded to the United States from Spain by treaty in 1819.

In 1867 Alaska was bought from the Russians and with this purchase the major land expansions stopped; thus in roughly 80 years (1781–1867) the United States had added 1,808,160,640 acres (2,825,251 sq. mi.) of land for $85,179,222 at an average cost of 50¢/acre.

It is surprising to the author that, after roughly 200 years of selling—and, more often, giving away land—the federal government still owned (as of June 30, 1960) 39.9% or 771,512,255.8 acre (1,205487.9 sq. mi.) of U.S. land. The difference or 1,501,894,464.2 acre (2,346,710.1 sq. mi.) is nonfederally owned land.

Table 3.2 shows the disposition of the public land by the United States. It is hard to imagine in a modern context, that in order to finance the construction of railroads (mostly the transcontinental routes) the federal government contributed 94.4 million acres, or just a little more than the entire state of Montana (145,600 sq. mi.). In 1851 the first U.S. land grant was made to the Illinois Central Railroad for 2,595,000 acres (about 4,055 sq. mi.), soon to be followed by other grants. There were some variations, but in general the railroads received alternating sections within 6 miles (or 10 miles, or 40 miles) for a maximum of six sections of land per mile of track constructed. Most of the land was sold as intended to settlers, but to this day the railroads retained most of the mineral rights (oil, gas, coal, etc.). In addition the railroads received, in most cases, a 100 ft, and in some cases a 200 ft right-of-way easement on either side of the centerline of the main track, as far as the main track ran.

Stories abound surrounding these dispositions. For example, there was Edward Gillette (1854–1936), after whom the town in Wyoming is named, who as the surveying chief of party, was in charge of locating the railroad alignment for several western railroads. He "often reflected that most people see the railroad as a big crew of laborers who are laying steel, getting drunk and raising hell generally. Only now and then did observant folk understand the significance of a small party of earnest men who came through the region, long before the track gangs, with surveying instruments, who worked swiftly and silently, said little or nothing to anybody, then passed on, leaving no trace of their coming other than a few stakes, driven here and there and there with no apparent system to the layman" (Holbrook, 1947). When under the mandate of "as soon as possible," he located 12 miles of railroad alignment in 48 hours; all this from horseback, without the benefit of modern aerial photos and GPS, "yet a large percentage of Gillette's original locations were never improved upon" (Holbrook, 1947).

Rights-of-way were also given to wagon roads (later highways), electric power companies, canals, irrigation ditches, etc., and in lieu of cash, veterans from the Civil War were paid with land. On the other hand, most states raised money, either by outright land sales or by scrip locations, which upon payment (mostly 50¢/acre), entitled the bearer to the land. (For details see: Title 43 of the United States Code, Public Lands, and Title 24 of the code, Mineral Lands and Mining.)

The disposition of public (federal) land to states, under Title 43, is noteworthy. Since the states retained their own individual property rights, the United States is the only major country in the world that does not have a uniform federal property law. Title 43 details what, how, and when each state was granted federal land. Especially in the western states, Section 16 and often also Section 36 of each township were set aside for schools or the financing of schools. In Oregon, the deeding of the school sections was subject to prior survey. In most cases the deed to the land included the surface and the underlying mineral rights. Again, in mostly western states, on admission to the Union, the state received grants for "Agricultural and Mechanical Colleges, State Universities and Common Schools," some states received 5% from the proceeds of the sale of public lands for the Common School Fund. Back to the railroads, it is not clear from the available data why an additional 37.1 million acres (58,000 sq. mi.) were deeded to the states for railroad construction.

Since the state of Texas has no U.S. public land, the state provided some large land grants—for example, 32,153,878 acres (50,240 sq. mi.) were deeded

UP = Union Pacific
CP = Central Pacific
NP = Northern Pacific
SP = Southern Pacific
AT = Atchison Topeka & Santa Fe
GN = Great Northern

U. S. Land Grants to Railroads

The Federal Government granted lands to railroads in alternate sections, retaining the sections between. The shaded routes show the approximate location of the land grants and are in proportion to the amount actually received. The only major railroad which did not receive any federal land was the Great Northern railroad.

FIGURE 42
Areas of U.S. land grants to railroads. (Adapted from Holbrook, Stewart H., 1947, *The Story of America Railroads.* Crown Publishers, New York.)

to the railroads at a rate of 16 sections of land per mile of track, provided at least 25 miles of track were constructed. Texas A&M University received 2,329,169 acres (3,639 sq. mi.) of land primarily in West Texas; mostly worthless land, until oil was discovered, and as of August 31, 1977, the University Trust fund had a value of $913,080,662. More than 42,500,000 acres (66,400 sq. mi.) of the state's public domain were granted to the Permanent Free School Fund, which has a value of $1,593,406,512 from the partial sale of land and mineral royalty revenues.

Supplemental Comments for Figure 43 to Figure 51

Figure 43 to Figure 45. Partial plat of township, T9N, R10E, Louisiana Principal Meridian. These three pages are an illustration of the longevity of property lines. The original plat, filed in 1829, shows French surveys surrounding Lake St. John and GLO surveys in the NW around Fletchers Lake. To be included on the plat, the ground surveys and possible recording must have been taken some time before the filing date. The modern maps (1976 and 1994) clearly show that the majority of the property lines and numbering of the lots have survived for at least 150 years. Ideally a modern satellite photograph should be included, since many of the survey lines are very visible from space, despite the lush vegetation.

Figure 46 to Figure 48. Spanish and Mexican land grants. In the south and southwestern part of the United States, many land subdivisions started as land grants during the roughly 1750–1850 time period. Many grants deeded by verbal conveyance were never patented by the American judicial system, which required written records.

Figure 46 probably represents the largest extent of land grants and in many cases only the grant names have survived to the present time. The Peralta Claim across the Arizona–New Mexico line represents one of the largest land fraud schemes in U.S. history.

Figure 47. Shows the maximum extent of land grants in Southern Colorado, most with rather vague boundaries. As can be seen, only a small fraction was ever adjudicated. The "Sangre de Cristo" grant later became Costilla County.

Figure 48. South central Colorado 1985 shows the "present" configuration. Starting in the NW is the Luis Maria Baca Grant No. 4, still shown as about a 100,000-acre block; since then (2004), most of the land (94,000 acres out of 100,000 acres) has been annexed into the Great Sand Dunes National Park. Costilla County is clearly shown with a distinct boundary. Until required in the future, possibly to settle a boundary dispute between counties, the boundary between Alamosa and Costilla counties is questionable. The "diagonal line" between an unnamed peak just south of Little Bear Peak and the banks of the Rio Grande River is described for Alamosa County in the Colorado Statutes (30-5-103) but there is no mention of it for the adjacent Costilla County (30-5-114). This may well become another legal case where a map has to take precedence over a written description. Not very visible on the enclosed map, but very visible on satellite photos is the scarcity of surface water, as evidenced by the Great Sand Dunes National Park. The purchase of the Luis Maria Baca Grant as an enlargement to the adjacent park yielded very little money for the ground surface, but most of the money was spent on subsurface water rights (total $31.28 million).

Figure 49. U.S. public land acquisitions. A snapshot in time with the clarity of hindsight. Many of the land transactions were vague at best, like the Louisiana Purchase (1803), which granted to the United States all the land westward to the continental divide, yet Lewis and Clark did not make their journey west for another 2 years. Ultimately the path of the continental divide was not finalized until well into the 20th century.

Figure 50. First surveys in Eastern Ohio. The layout of the "Old Seven Ranges," the "Ohio Company Purchase," and the "Donation Tract." After the "Geographers Line" was laid out westward from the intersection of the Pennsylvania state line and the north shore of the Ohio River, meridians were surveyed south until they intersected the Ohio River. Numbering of the townships progressed northward from the river with the numbering of the sections as shown. The Ohio Company Purchase and the Donation Tract followed the established pattern of the first survey. Very little of the original survey lines are still visible today.

- East Liverpool, Ohio, is located about 5 miles west of the initial point, on the Ohio River;
- Marietta, Ohio, is located about T-1, R-8, on the Ohio and Little Muskingum River;
- Athens, Ohio, is located about T-11, R-14, on the Hocking River and the former Hocking Canal.

Figure 51. Original Ohio land subdivisions. The present state of Ohio served as a proving ground for subdividing federal lands. As a general statement, the expectations were high, the end results mostly disappointing. Even though shown on Figure 51, most tract boundaries were not as clear as shown on the map because many lines were either never or only poorly surveyed; true meridians (north–south lines) and true parallels (east–west lines) were especially difficult to survey. Unlike the subsequent surveys to the west of Ohio where the GLO rectangular pattern is still very visible, on the ground and especially from the air, by contrast, the original survey patterns in many places in Ohio have mostly disappeared and only fragments, such as corners, etc., remain.

FIGURE 43

Original (?) French tract surveys, surrounded by rectangular GLO surveys. GLO plat for T9N, RIOE, Louisiana Principal Meridian, filed 1829.

FIGURE 44
Modern (1976 and 1994) maps of T9N, R10E, Louisiana Principal Meridian.

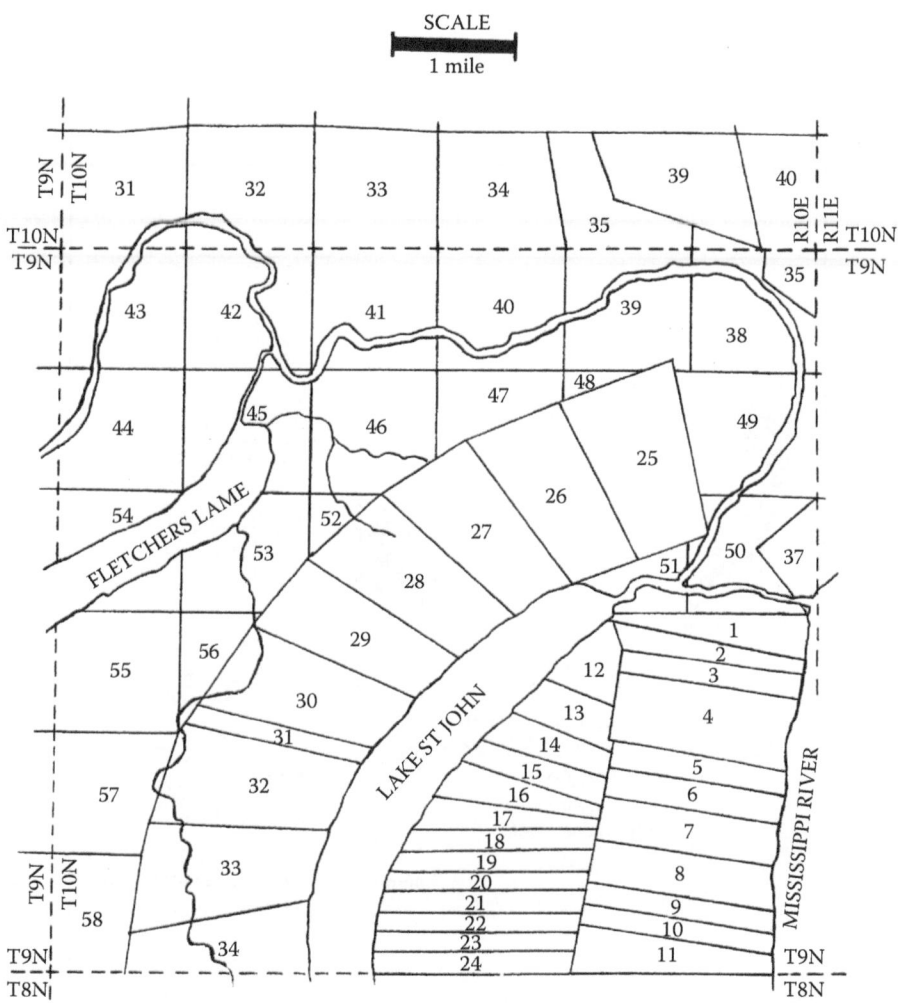

Abstract of Property Lines in T9N, R10E, Louisiana PM
(based on 1976 & 1994 USGS topo maps)

There has been virtually no change in property line alignments since the original GLO plat (see Figures 43 and 44) was drawn in 1829. Undoubtedly there are now, but too small to draw, many private properties for fishing or weekend cabins along the west shore of Lake St John, as evidenced by the many house symbols and fishing piers shown on the maps (Figure 44).

FIGURE 45
Abstract of Property lines in T9N, RIOE, Louisiana PM.

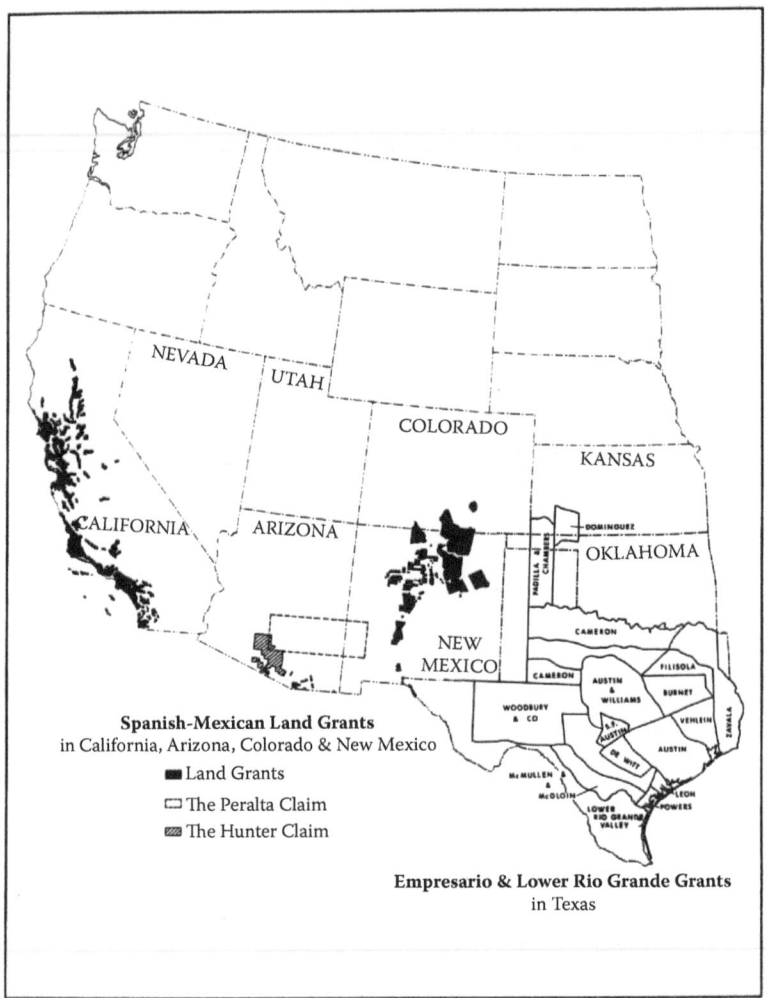

FIGURE 46
Spanish and Mexican land grants. © 1989 University of Oklahoma Press.

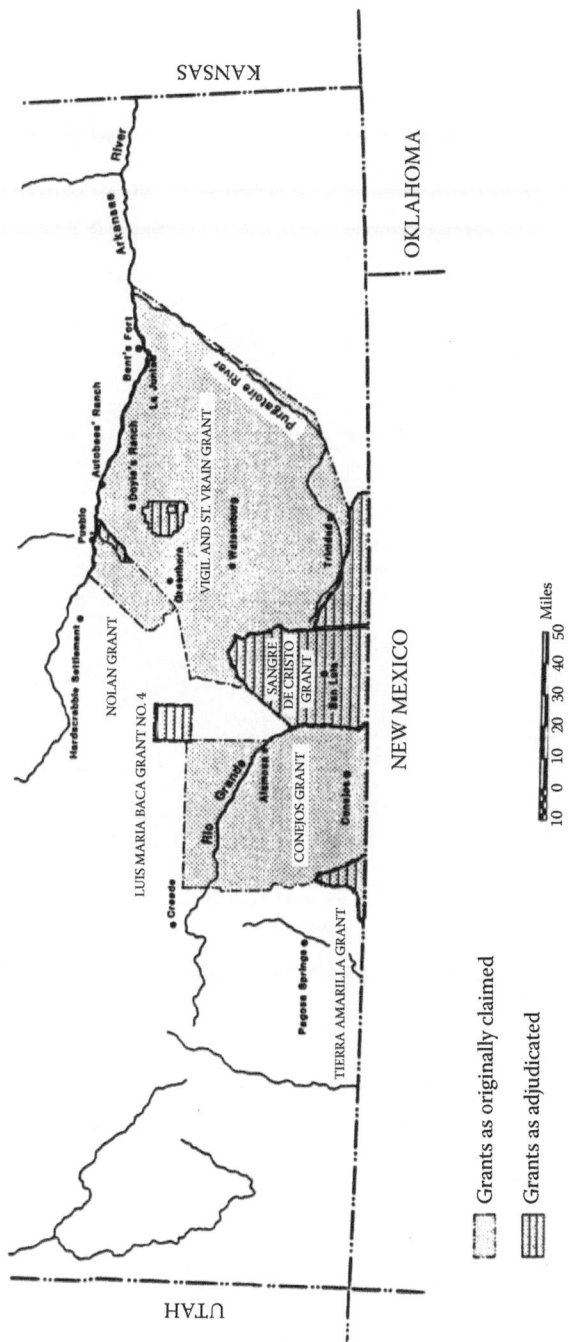

FIGURE 47

Mexican (Spanish) land grants in southern Colorado. © 1994 University of Oklahoma Press.

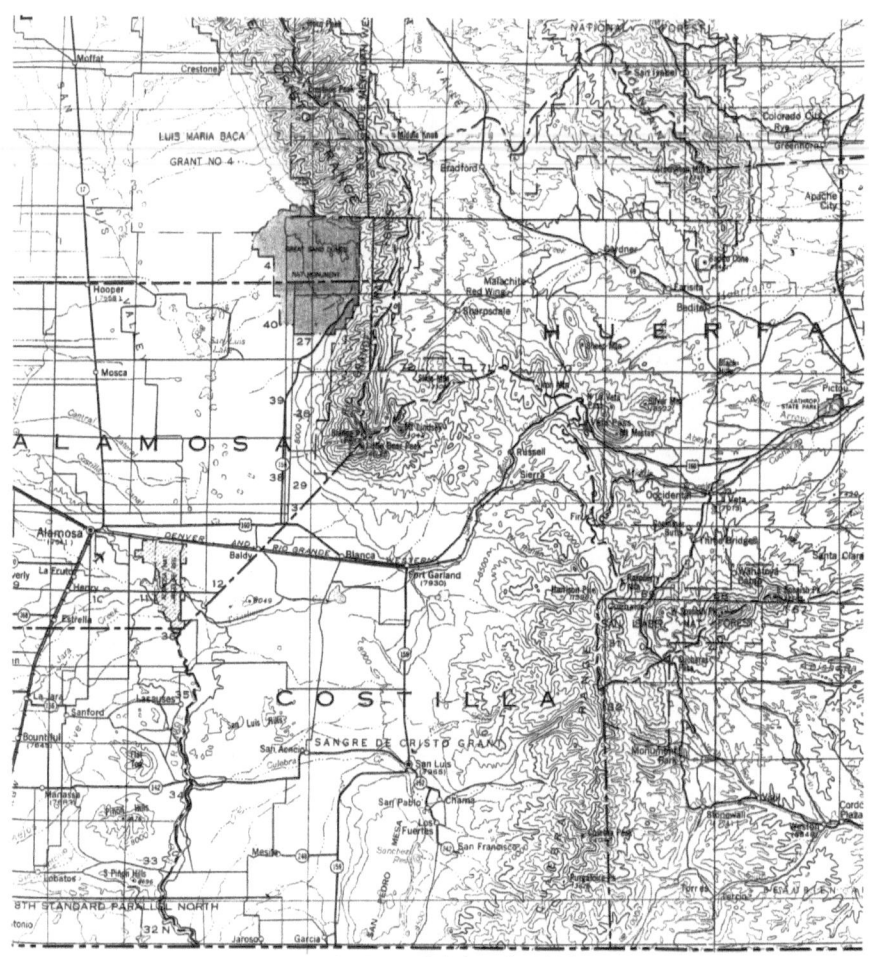

South -Central Colorado 1985
Scale 1: 500,000
Note: The boundaries of the Luis Maria Baca Grant No. 4 and the Sangre de Cristo Grant.

FIGURE 48
The legacy of Spanish and Mexican land grants in southern Colorado.

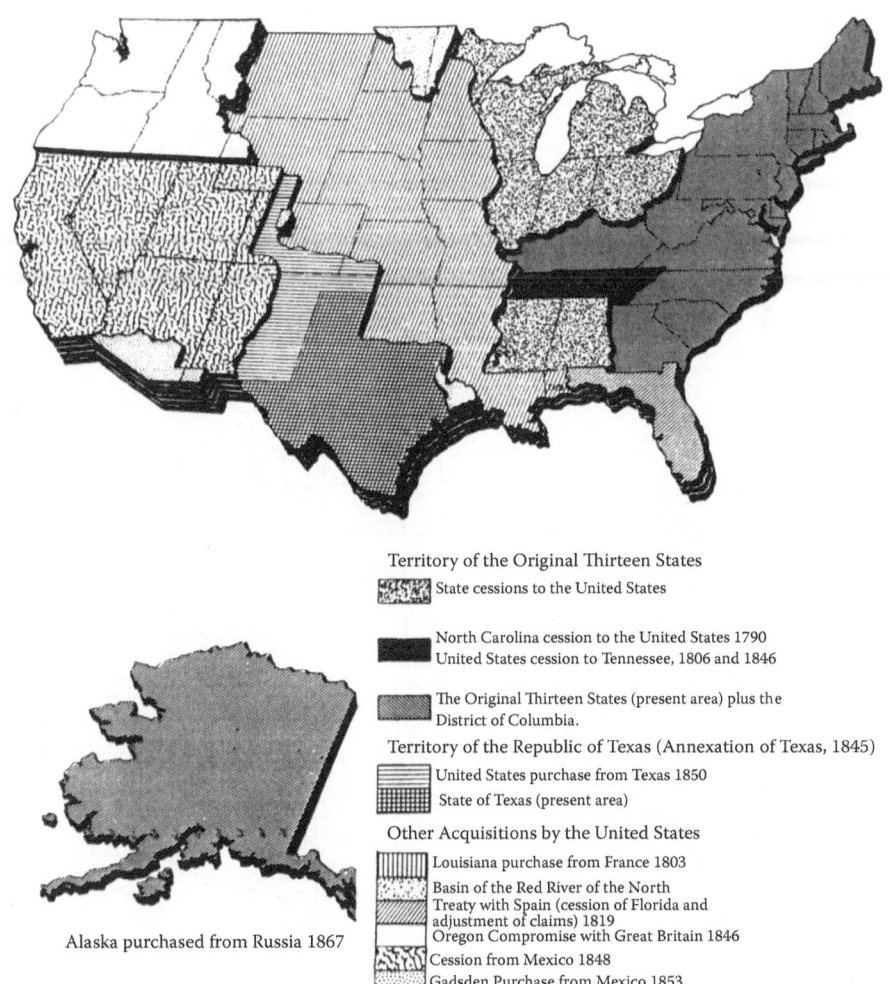

Territory of the Original Thirteen States

State cessions to the United States

North Carolina cession to the United States 1790
United States cession to Tennessee, 1806 and 1846

The Original Thirteen States (present area) plus the District of Columbia.

Territory of the Republic of Texas (Annexation of Texas, 1845)

United States purchase from Texas 1850

State of Texas (present area)

Other Acquisitions by the United States

Louisiana purchase from France 1803

Basin of the Red River of the North
Treaty with Spain (cession of Florida and adjustment of claims) 1819
Oregon Compromise with Great Britain 1846

Cession from Mexico 1848
Gadsden Purchase from Mexico 1853

Alaska purchased from Russia 1867

Source: Bureau of Land Management

FIGURE 49
U.S. public land acquisitions.

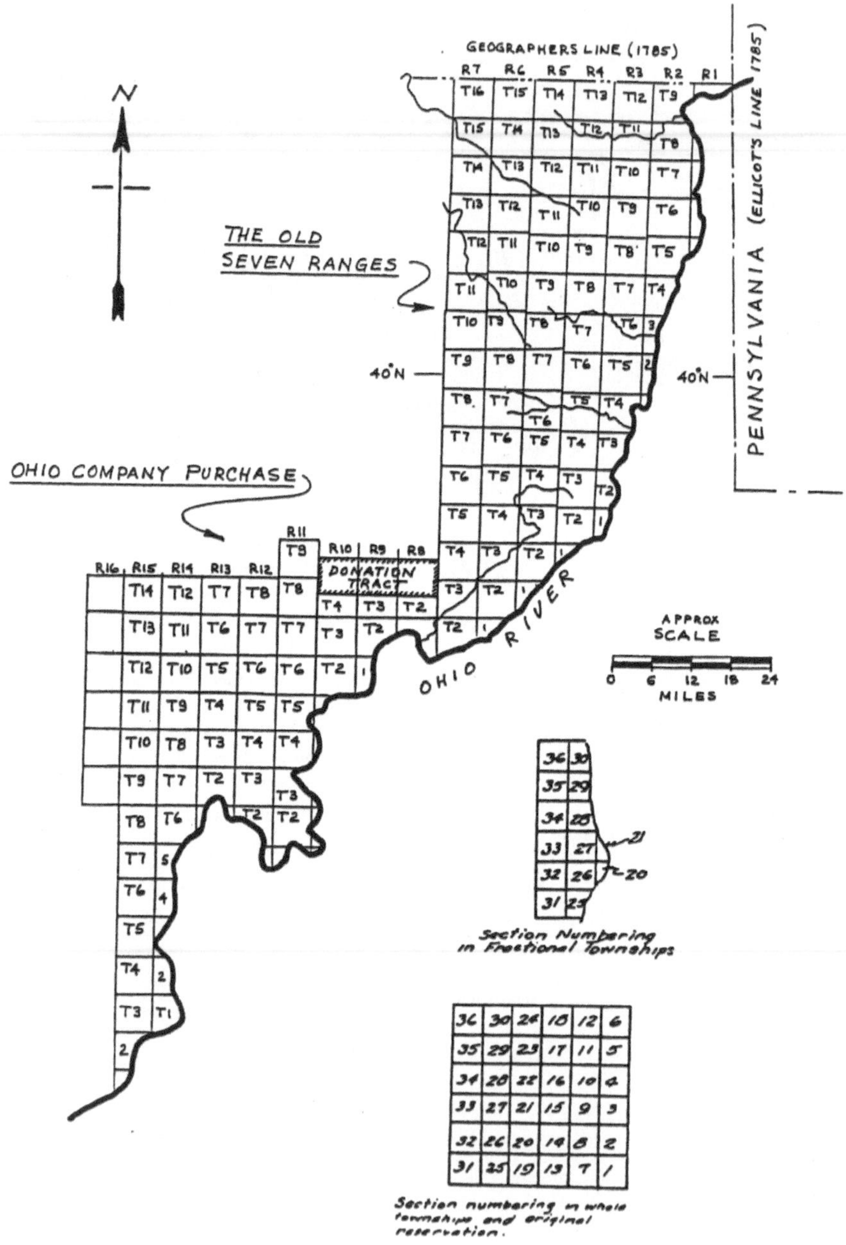

FIGURE 50
First surveys in eastern Ohio.

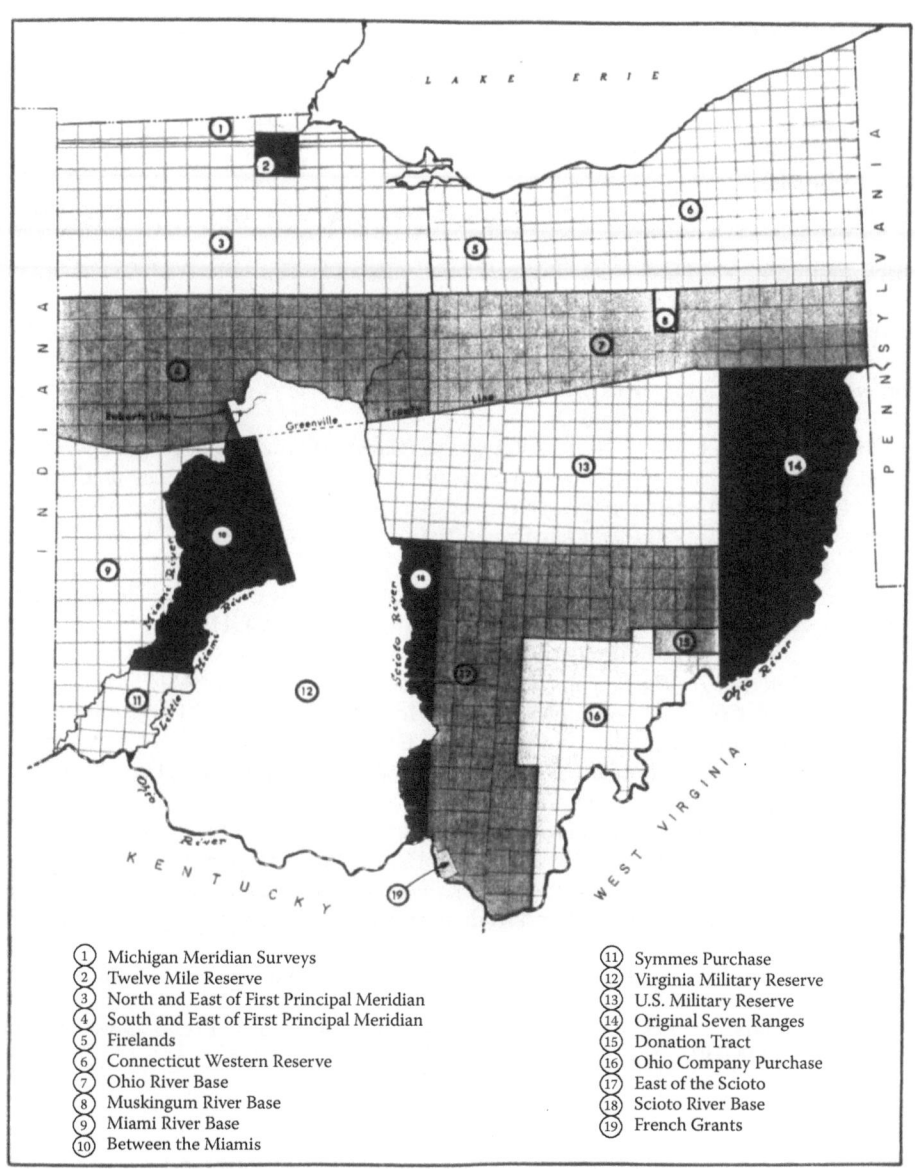

1. Michigan Meridian Surveys
2. Twelve Mile Reserve
3. North and East of First Principal Meridian
4. South and East of First Principal Meridian
5. Firelands
6. Connecticut Western Reserve
7. Ohio River Base
8. Muskingum River Base
9. Miami River Base
10. Between the Miamis
11. Symmes Purchase
12. Virginia Military Reserve
13. U.S. Military Reserve
14. Original Seven Ranges
15. Donation Tract
16. Ohio Company Purchase
17. East of the Scioto
18. Scioto River Base
19. French Grants

FIGURE 51
Original Ohio land subdivisions.

Selected Bibliography

ACSM, ASCE, 2005; *Definitions of Surveying and Associated terms*, The American Congress on Surveying and Mapping and the American Society of Civil Engineers. Gaithersburg, Maryland.

Bartlett, Richard A., 1962, *Great Surveys of the American West*, University of Oklahoma Press, Norman.

Beck, Warren A. and Hasse, Y.D., 1989, *Historical Atlas of the American West*, University of Oklahoma Press, Norman.

Brown, Curtis M., and Eldridge, W., 1962, *Evidence and Procedures for Boundary Location*, John Wiley & Sons, New York.

Brown, Curtis M., Landgraf, F.H., and Uzes, F.D., 1969, *Boundary Control and Legal Principles*, John Wiley & Sons, New York.

Holbrook, Stewart H., 1947, *The Story of American Railroads*, Crown Publishing, New York.

Noel, Thomas J., Mahoney, P.F., and Stevens, R.E, 1994, *Historical Atlas of Colorado*, University of Oklahoma Press, Norman.

Peters, William E., 1930, *Ohio Lands and Their History*, reprinted 1979, Arno Press, New York.

Sherman, Christopher E., 1925, Original Ohio Land Subdivision, Ohio Topographic Survey, Ohio Division of Geological Survey, Columbus, OH.

Stewart, Lowell O., 1935, *Public Land Surveys*, Collegiate Press, Inc., Ames, Iowa. Reprinted 1976, *Public Land Surveys: History, Instructions and Methods*, Meyers Printing Co., Minneapolis, MN Reprinted 1979, Public Land Surveys—Development of Public Land Law in the United States, Arno Press, New York.

United States Code Annotated, 1964; Title 43 Public Lands, Section 1 to 1383, West Publishing Co., Eagan, Minnesota.

White, C. Albert, 1996, *Initial Points of the Rectangular Survey System*, Professional Land Surveyors of Colorado, The Publishing House, Westminster, Colorado.

White, C. Albert, 1983, *A History of the Rectangular Survey System*, U.S. Department of the Interior, Bureau of Land Management, U.S. Government Printing Office, Washington D.C.

White, Richard, 1991, *A History of the American West*, University of Oklahoma Press, Norman.

Wyckoff, William, 1999; Creating Colorado, Yale University Press, New Haven, CT 06520-9040.

Mining Claims and Related Items

The material below is a generalized overview relating to mining claims and related activities. Anybody who is considering professional work in this area should first consult all applicable federal and state laws and their amendments.

In principle, the concept of a mining claim provides an individual the right to assert the right of possession for the purpose of extracting a discovered mineral deposit on public domain lands. As originally written in the U.S. General Mining Law of 1872, mining claims provide legal protection to an individual, with respect to place and time, after the discovery of a mineral deposit, and in its strictest sense does not address landownership.

Several corollaries follow the paragraph above.

- The U.S. General Mining Law of 1872 and its subsequent amendments address mining on federal lands only; there are additional state and local mining district laws to be considered. Mining claims can still be staked in unreserved and unappropriated federal public lands in Alaska, Arizona, Arkansas, California, Colorado, Florida, Idaho, Louisiana, Mississippi, Montana, Nebraska, Nevada, New Mexico, North Dakota, Oregon, South Dakota, Utah, Washington, and Wyoming. For dimensions of mining claims, see Figure 52.

- Areas closed to mineral prospecting are National Parks and Monuments, wilderness areas, and wildlife refuges, Indian and military reservations, and public lands withdrawn by congressional or presidential acts.

- Prospecting and mining on lands in states that reserved mineral rights to themselves before joining the Union are governed by state and local laws and ordinances.

- Prospecting and the extraction of minerals on private land is considered trespassing unless permission/compensation by the landowners is first obtained.

For the purpose of the Mining Law, minerals can be classified as locatable, salable, and leasable:

> Locatable minerals: Locatable mineral deposits can be claimed on federal public lands, under the General Mining Law of 1872 and subsequent amendments.
>
>> *Metallic minerals*—gold, silver, lead, zinc, copper, mercury, antimony, etc.
>>
>> *Non-metallic minerals*—mica, uranium, fluorspar, asbestos, etc.

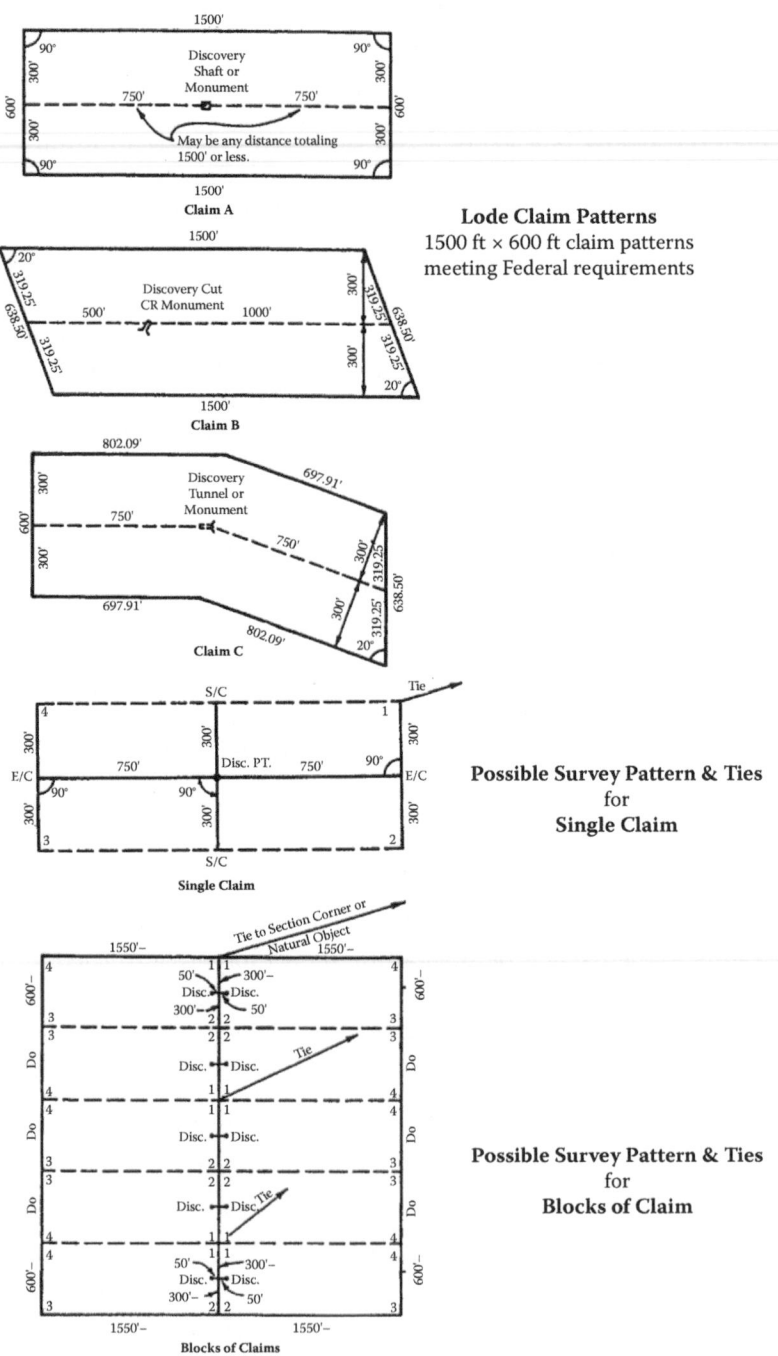

Lode Claim Patterns

1500 ft × 600 ft claim patterns meeting Federal requirements

Possible Survey Pattern & Ties
for
Single Claim

Possible Survey Pattern & Ties
for
Blocks of Claim

FIGURE 52
Possible dimensions of mining claims.

Salable minerals: Salable mineral deposits can NOT be claimed under the General Mining Law of 1872, but may be purchased from the federal government, under the Materials Act of 1947 and subsequent amendments.

Salable minerals—sand, gravel, stone, pumice, clay, etc.

Leasable minerals: Leasable mineral deposits can NOT be claimed under the General Mining Law of 1872, but may be leased from the federal government under the leasing act of 1920 and subsequent amendments.

Leasable minerals—oil, natural gas, coal, oil shale, geothermal resources, native asphalt, bituminous rock, borax, phosphate, sodium, and sulfur in Louisiana and New Mexico.

Where permitted by federal law, there are four types of mining claims, namely, lode claims, placer claims, mill sites, and tunnel sites.

Lode Claims

A lode claim can be staked when there is evidence of a bona fide mineral deposit, and the mineral must be firmly contained or embedded in solid rock (such as a vein, replacement or disseminated deposits, etc.). A claim is valid only after a valuable mineral deposit has been discovered and it meets the "prudent man and marketability test." The test requires that a "person of ordinary prudence" would be justified in further expenditure of his labor and means, with a reasonable prospect of developing a viable mining operation (i.e., that minerals can be extracted, removed, and marketed at a profit). The discovery can be on the surface or can be made underground. The date of discovery determines the legal rights of the claimants as well as the seniority of the claim.

A lode claim cannot exceed a parallelogram 1,500 ft in length by 600 ft in width and parallel ends (see Figure 52). As an indispensable condition of discovery, the discovery point and the boundary of the claim must be marked on the ground (monumented) within a prescribed period of time and recorded with the proper state agency(s). In staking out a claim, it is advisable to make the sides shorter than the 600 ft × 1500 ft, so that its dimensions cannot be challenged for oversizing by a resurvey for oversizing. The boundary of a lode claim is described by a metes and bounds survey description with a tie to a GLO monument (no more than 2 miles distant) or to a prominent natural object (mountain peak, fork in river, etc.). It is possible to amend an original claim location for a "better" alignment with a vein, etc., for extralateral mining rights (see Figure 53 and Figure 54). Prior to the moratorium of 1994, it was possible to patent a lode mining claim.

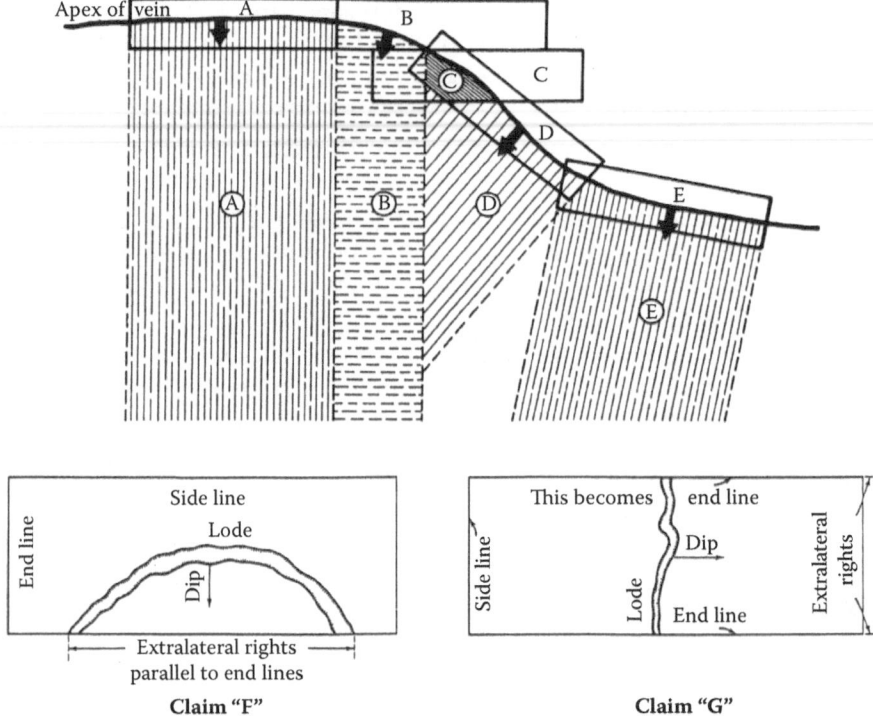

Claim "A" – Ideal claim; strike parallel to side line of claim; extralateral rights "downdip".
Claim "B" – Vein crosses side line of claim and extralateral right stops.
Claim "C" – Vein leaves the claim through both side lines, there are NO extralateral rights.
Claim "D" – Claim "D" is junior to claim "C", but senior to claim "E".
Claim "E" – The vein passes through the end lines, but is junior to claim "D".
Claim "F" – Strike of vein passes through both end lines (assuming discovery point is at center
  of claim), extralateral rights are as shown.
Claim "G" – The original side lines can become end lines.

FIGURE 53
Mineral rights associated with a lode claim.

Assessment Work

As originally written, $100/year of physical assessment work had to be done on each mining claim "in order to hold the possessory right to a lode or placer claim" and many acres of federal land were needlessly damaged. On October 5, 1992, Public Laws 102-381 and 103-66 and October 21, 1998, amended the original mining laws; a claimant with 10 or fewer claims is still allowed to perform assessment work; however, all other claimants must pay an annual fee of $100/claim or site to the BLM by each August 31. There is no limit on the number of renewals. In lieu of a payment, it is possible to obtain a waiver. Failure to pay the annual fee will null and void the claim(s) and forfeit all rights.

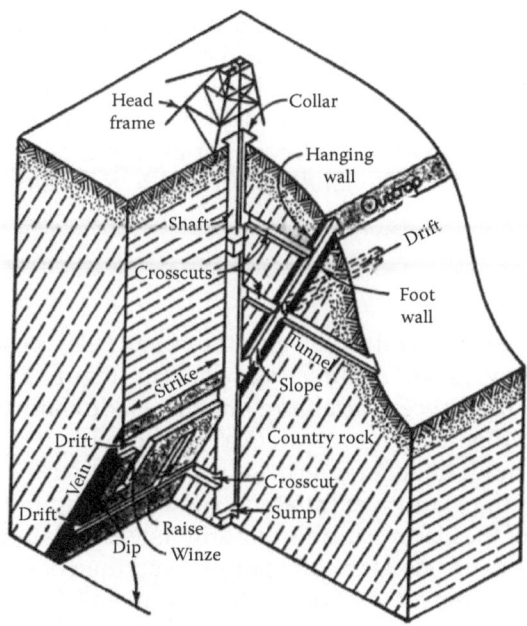

Mining Terminology

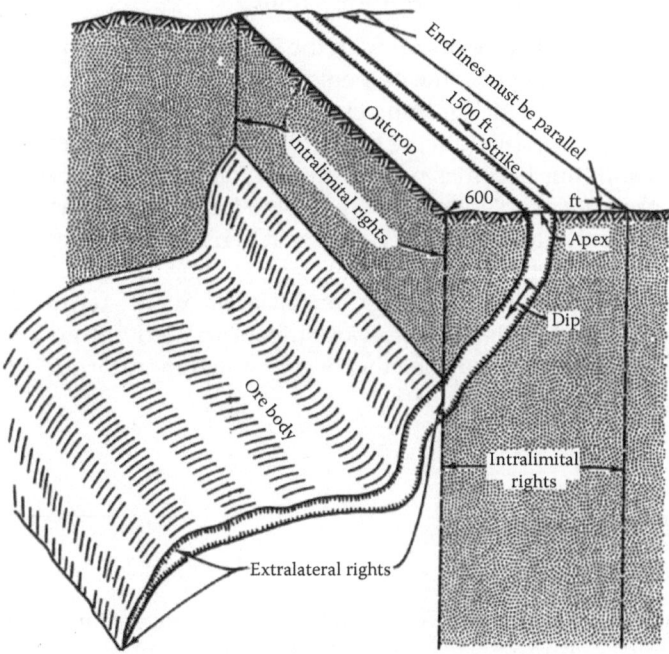

Extralateral & Intralimital Rights on a Vein

FIGURE 54

Drill holes, geological, geochemical, and geophysical surveys can qualify for assessment work; on the other hand, road and bridge work do not qualify.

There are many local, state, and federal regulations to protect the surface resources during exploration activities. Both the extent (area) and intensity of the activities have to be considered. For example, casual activities cause negligible disturbance to the land and do not involve earth-moving equipment or explosives.

Extralateral Right

The principle of extralateral rights and the "apex rule" pose a unique problem to both surveyors and mining engineers (see Figure 53 and Figure 54).

The relationship between the alignment of the lode claim and the "outcropping or apex" (either actual or theoretical) of the vein on the ground control the limits of mining. In general terms it is possible to mine "down dip" past the side lines of the claim = extralateral rights; mining must stop vertically at the endlines of the claim = intralimital rights. "Extralateral rights exit only when the end lines are substantially parallel and then only when the apex of the vein passes through at least one of the end lines."

Mining Claim versus Patented Mining Claim

A mining claim provides the owner with the right of discovery (place and time) of a valuable mineral deposit; the ownership of the land stays with the federal government. With a patented mining claim, an actual mining venture is or has taken place, and ownership (title) of the land has been granted to an individual (or company) and thus becomes private property; it is also not necessary to pay annual fees or perform assessment work. A step for a lode or placer claim to go to patent requires that a mineral surveyor must perform the actual claim survey and the monuments set during the initial claim staking control.

Following the Federal Land Policy and Management Act of 1976, public lands are to remain under the stewardship of the federal government. In addition, since 1994 there has been a congressional moratorium in effect for granting patents on claims or sites.

Placer Claim

Historically, a placer claim is an unconsolidated concentration of precious minerals (gold, platinum, etc.) which have been mechanically transported and then deposited by gravity differences; by Congressional acts and through judicial interpretation a placer claim can be nonmetallic bedded or layered deposits, such as gypsum or high calcium limestone, etc. A placer claim is a maximum of 20 acres per person (160 acre maximum) and described by the legal subdivision of the public land survey. Prior to the moratorium of 1994, it was possible to patent a placer claim and the boundary determined the vertical extent of the claim.

Mill Site

"A mill site is required to be used or occupied distinctly and explicitly for mining or mining purposes," usually covering an area of 5 acres and described by the legal subdivision of the public land survey. A mill site may not be patented.

Tunnel (Adit) Site

A tunnel site is located where a tunnel or adit is run to develop a vein or lode; 3,000 ft by 3,000 ft. A tunnel site may not be patented and failure to work the tunnel for 6 month constitutes abandonment.

CAUTION: When entering the world of mining claims, a neophyte quickly learns that he or she has entered the realm of boundless optimism (i.e., tomorrow we will strike the big one, etc). A cursory check of the Internet will show that there are untold opportunities to be had from claims to gold mines and any other mining venture driven by the then current high value of mostly metallic commodities. Undoubtedly there are still many "golden" opportunities but a little caution and some common sense is advised. Not necessarily in that order:

- Mining claims can only be located on federal land open to mineral entry; it is the claimants' responsibility to check ownership (federal) and if the land is open to mineral claims.
- Since the Federal Land Policy and Management Act of 1976 and the congressional moratorium in 1994, all federal lands will remain under the stewardship of the federal government. From this, it follows:
- *The federal government retains ownership of the land;* the claimant only is granted the right to extract valuable minerals, i.e., the federal government will no longer deed land with a patented mining claim as in the past.

  Without any valuable minerals the claimant's mining claim(s) becomes null and void; the claim can NOT be used for private non-mining purposes, i.e., mountain cabin, etc. Another corollary is that at the end of the mining operation the land reverts back to the federal government (actually the land was just leased), and it has to be restored by the claimant to the condition prior to mining activities.

The claimant has to allow public access to and across the claim, i.e., the claim cannot be fenced off.

The public cannot cross private land to reach public land, unless the private landowner has granted permission.

It is not necessary to employ a professional land surveyor or a mineral surveyor to survey and describe the metes and bounds boundary of the claim or for a placer claim describing by aliquot parts and lots with the U.S. Public Land Survey System.

Upon discovery, it is necessary to post a location notice on site, file the notice in the appropriate county, and also file within 90 days with the Bureau of Land Management (BLM, which also assigns an ORMC claim number). It is necessary to observe the laws of some original mining districts, as well as state and federal (especially consult Title 43: Public Lands, part 3832) laws. The unauthorized use of a mining claim can become a very serious problem to the claimant.

Especially in the past, physical assessment work on a mining claim often was brutally destructive to the environment. Even though still permitted, most claimants prefer to pay an *annual fee* to the BLM. Non-payment will be considered abandonment of the claim(s).

Some Mining Terminology

Adit: A horizontal entry into a mine with one opening to the surface, commonly and erroneously called a tunnel. (A tunnel is open at both ends!)

Apex: The top (usually outcropping on the surface) of a vein or lode.

Assay: A test to determine the amount and type of minerals in a given sample. May be made with a miniature smelting process in the laboratory, called a "fire assay" or by the use of chemicals, called a chemical or wet assay.

Collar: The timbering or concrete works around the entrance of a shaft or winze (see Figure 54).

Contact: The meeting of two geological formations, such as country rock and a vein.

Contact vein: A vein along the contact

Country rock: Non-mineralized rock surrounding a vein or lode (see Figure 54).

Cribbing: A wall of light (wood) timbering between heavy supports.

Crosscut: A horizontal passage through country rock, connecting one working with another (see Figure 54).

Cut: An open working driven into a hillside to expose underlying rock.

Dip: The downward slope angle of a formation (vein, bed, etc.). True dip is measured perpendicular to the strike, for example: the dip is 34° SE'ly; strike and dip frequently go together, such as the vein has a strike N36°E, dip 47° W'ly.

Drift: A horizontal working in the vein, i.e., a working following the strike of the vein (see Figure 54).

Face: The last working at the end of an adit, drift, crosscut or cut.

Fault: A fracture plane or fracture zone in the rock. Originally continuous ore bodies may be cut by faults and displaced. Faults also may provide a conduit for mineralized fluids and subsequent ore deposition.

Float: A piece of ore detached from a vein or lode lying loose, not in place.

Foot wall: The rock surface of a vein or fault, under foot of a person walking in a drift; the opposite of the hanging wall (see Figure 54).

Grizzly: A grating usually made from mine rail or heavy steel bars for the purpose of separating different rock sizes.

Grub stake: The financing of a prospector for a share in his (future) findings.

Hanging wall: The rock surface of a vein or fault, hanging above a person walking in a drift. The opposite of the foot wall (see Figure 54).

Level: A horizontal working.

Muck: The broken non-mineralized rock in a mine. A machine or person to handle the muck is called a "mucker."

Open cut: A trench made in the open for the purpose of exploration.

Placer: A mineral deposit of unconsolidated particles; for example, gold flakes in a sand bank.

Raise: A working driven upward from below (usually in the vein) (see Figure 54).

Royalty: A percentage of the earnings or product paid to the owner; a mineral severance tax is usually paid to the local, state, and federal governments.

Shaft:	A vertical, or approximately vertical, opening from the surface of the mine working (see Figure 54).
Skip:	An ore bucket used to hoist ore or muck up the shaft to the surface.
Slope (or incline):	An inclined shaft that follows the vein. It is usually down dip, i.e., it is parallel to the dip.
Smeltering:	The reduction of ore or ore concentrates in a furnace into metal.
Strike:	The bearing of a horizontal line in a geological feature (formation, vein, bed, fault, etc.). For example: the strike of a vein is N20°E; strike and dip frequently go together, such as a vein has a strike N36°E, dip 47° NW'ly.
Stope:	An irregular opening in the mine where ore is mined.
Sump:	A low place in a mine for collecting mine water.
Tailings:	Waste rock from a mine or a mill.
Tunnel:	A horizontal passage open at both ends.
Winze:	A working driven downward from a level, usually a vein (see Figure 54).
Working:	Any excavation, tunnel or passage, made by mining.

Selected Bibliography

Brown, Curtis M., Robillard, W.G., and Wilson, D.A., 1995; *Brown's Boundary Control and Legal Principles*, 4th Ed. John Wiley & Sons, New York.

Electronic Code to Federal Regulations; Title 43: Public Lands: Interior, Part 3832. Address:http://ecfr.gpoaccess.gov.

U.S. Department of the Interior, Bureau of Land Management; Staking a mining claim on federal lands, date unknown.

> 1980; Mineral survey procedure guide, GPO 857-030.
>
> 1996; Mining claims and sites on federal lands, GPO 1996-776-268.
>
> 2007; Locating Mining Claims, Information Guide; Oregon/Washington State Office.

Water Laws

Water Rights

Water rights (surface and groundwater), mineral rights, etc., are different from property rights and often are completely separate from each other. Each state has jurisdiction over its water, but in general in the U.S. two types of water rights exist. They are:

1. Riparian water rights
2. Prior appropriation water rights.

<u>Riparian water rights.</u> In states where water is abundant (eastern U.S. ±) the riparian doctrine is applied. The principle is that the water belongs to the public (under state control) and, for example, the abutting property owner has the use of the water. In case of water shortages, the "equal fairness" rule may be applied, where everybody will be able to draw an equal (but reduced) share from the available water pool. In some areas the "most beneficial" rule had to be used, for example, first call for domestic water, second agricultural, etc.

<u>Prior appropriation water rights.</u> In states where water is a limited resource (western U.S., more or less), the prior appropriation water doctrine is applied. The principle states that the water (both surface and groundwater) is NOT a public, but a private resource for beneficial use and that the first claim or call IN TIME has absolute priority over everybody else, i.e., "first in time, first in right." Stated another way, unless a landowner has the water right, he can NOT use the water flowing past or under his property. In case of a water shortage, the senior right (for example, 1871) has priority over the junior right (1905), and the junior may receive no water at all. Water rights are independent of land property rights and water can be bought, sold, or traded like any tangible property. The prior appropriation rights are held only as long as proper beneficial use is continued. For arid lands, the water rights may be worth more than the land.

An application for prior appropriation water rights must start with an application to the state water engineer who determines the amount required, the "availability of water in a drainage basin." After an affirmative ruling, the next step is a request to the state water court, which must rule on the description of the water source, the amount of water claimed, the use of the water, and the date of initiation (i.e., the beneficial use). A favorable ruling then also sets a time of adjudication. Undoubtedly many homeowners give little thought to whether their irrigation well in the backyard is adjudicated or not, but rightfully it should be.

Several other concepts are also applicable:

Use of water. The water may only be used in the drainage basin or aquifer from which it is drawn, but not consumed or removed. For example, legal uses are domestic use, returning the water through the septic or sewer; agricultural use, returning the surface water to the groundwater, etc. Illegal uses would be the exporting or consumption of the water in long distance slurry pipelines or shipping beverages outside the drainage basin.

Water quality. In recent years water quality has become a major problem. The principle is simple: the water must be returned in the same quality (chemical composition, metal, and salt content, temperature, etc.) as originally taken in. State and federal regulations address and continuously add to this topic.

Point of diversion. Generally water rights not only specify the amount of water but also the point where it can be drawn (i.e., point of diversion). For example, the town of Golden had (has) considerable problems because water was drawn from Clear Creek above the town, and then was discharged through the sewer system in North Denver, about 15.5 miles below Golden. Owners of water rights between Golden and the discharge of sewage treatment plant were therefore deprived of their water. Thus, in lean water years, Golden may be forced to pump millions of gallons of water back upstream to satisfy some water rights at their point of diversion. This would also involve the building of a new water pipe as well as considerable pumping costs.

Adjudication of water wells. In Colorado the location and capacity of a water well, for either domestic or industrial use, must be legally filed and recorded with a water court (i.e., adjudicated). Adjudication establishes the legal time element for prior water appropriation. The state is divided into drainage basins and the state water engineer has to approve a drilling permit and water withdrawal rates. See previous description of *prior appropriation water rights*.

Irrigation ditch easements. In Colorado irrigation ditch easements are protected by the state constitution, and a ditch has the dominant right over the rights of a property owner. A property owner does not have the right to change the alignment or configuration of a ditch. Irrigation ditch easements are probably not filed and recorded.

Water Laws

Abstracted from "Groundwater," by Earl Cornish, *Colorado School of Mines Mineral Industries Bulletin*, Vol. 10, No. 4, July 1967.

Throughout the United States four doctrines or rules form the legal basis for groundwater disputes. They are:

1. The English "common law" doctrine or "absolute ownership" rule
2. The American rule or "reasonable use" doctrine
3. The California doctrine or "correlative rights" rule
4. The doctrine of prior appropriation or "prior rights" rule

We will first discuss the nature of each doctrine and then use a hypothetical situation to add clarification where needed.

The English Common Law Doctrine or Absolute Ownership Rule

The first of these doctrines, the Absolute Ownership rule, goes back a long way. It evolved out of a legal dispute concerning groundwater withdrawal between two individuals in England in 1821. The suit was brought to court by a Mr. Action, a cotton mill owner, who had been pumping water from a particular aquifer for many years. He wanted damages assessed against a Mr. Blundell, who had sunk two coal pits in the area of his cotton mill. These coal pits were below the water table level in the area and consequently were causing Mr. Action's well to dry up. The court ruled in favor of Mr. Blundell, however, and refused damages to Mr. Action. The judge ruled: "Whose the soil is, his it is from the heavens to the depths of the earth."

This then is the English rule. It says in effect that the owner of a piece of property has the absolute right to all the water that passes or is stored beneath his property. Until shortages in groundwater began to appear, the English rule was widely accepted in the United States. But since it was not based on an adequate understanding of groundwater withdrawal and replenishment, it obviously was not suited for equitably solving disputes in areas of severe depletion such as our arid West.

The unsuitability of this rule can be seen from the not uncommon situation of several farmers withdrawing water from the same aquifer where one of the farmers has land located at the most favorable spot along the reservoir (see Figure 55). It is obvious that in this situation during times of drought, when the water table is sinking within the aquifer, some controls would have to be placed on the amount of water each farmer could withdraw. For if this were not done, Farmer D could force the drying up of all the wells on the adjoining property by using some of the water on his own land and selling some to Farmer E. Because of this situation and others like it, the English rule has not been very popular in this country, especially in arid regions.

The American Rule or Reasonable-Use Doctrine

The American rule grew out of dissatisfaction with the English rule of absolute ownership. However, it is virtually the same doctrine except for two

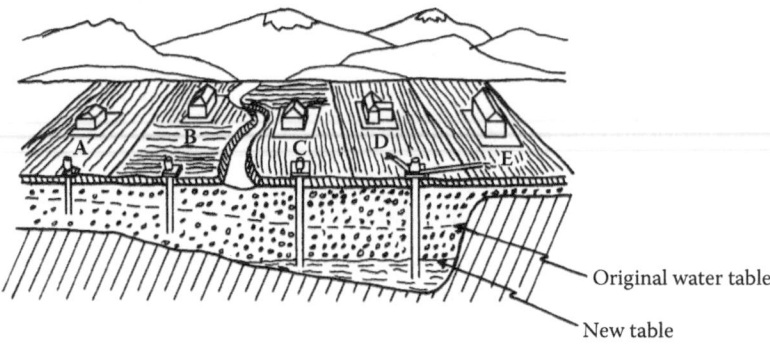

Original water table

New table

FIGURE 55

particulars. Specifically, these state that it is not permissible to divert or sell groundwater outside the basin from which it was drawn unless injury does not result to neighboring well owners, and it is also not permissible to maliciously use the water. This rule is perhaps in some ways even less efficient than the English system since it affords so many areas of arbitrary interpretation. It developed out of a dispute involving the city of Brooklyn and a farmer in Kings County, New York, during 1885. The farmer had for years been irrigating his land with groundwater. Subsequently, the city of Brooklyn sank a well near his land and constructed a pumping station to export water into the city. This action had the effect of lowering the water table and the farmer's well began to go dry. His crops failed and he appealed to the courts. The judgment was in his favor with the court declaring that the city of Brooklyn was making an unreasonable use of the underground water. Reasonable use did not include the withdrawal of groundwater for uses not associated with the land from which it was taken. The definition of reasonable use of water must depend upon the courts for interpretation.

The American rule is considered in many quarters to be a considerable improvement in justice over the English system, especially in the eastern part of the United States, but in the arid western regions stronger laws had to be developed. The California doctrine or correlative-rights rule is just such a measure.

The California Doctrine or Correlative-Rights Rule

The correlative-rights rule came into effect as a result of a decision handed down by the California supreme court in 1903. The dispute involving this decision had first been decided in a lower court under the reasonable use theory. However, when the case came up for trial the second time, the judge went one step beyond the rule of reasonable use and introduced a new doctrine. He stated that each overlying landowner had an equal and "correlative right" to develop and use the water beneath the land. By correlative he meant that each right was mutually related and dependent upon other water rights in the basin. This law then states that reasonable use of a groundwater reservoir by any individual is *relative* to the other water users in the basin. When a surplus of water exists, the California rule permits the sale or exploitation of the amount declared to be surplus.

Again referring to the situation of several neighboring farmers drawing water from the same reservoir, we can quickly see how this rule would apply. Figure 56 demonstrates what would happen under drought conditions if the reasonable use rule was the only law in effect. Here Farmer D is using the water he is continuing to withdraw from the reservoir to irrigate his own land; no one can stop him should the courts decide that this is a reasonable use of the water. In Figure 57 we can see that under the same drought conditions, even though the water table is lowered, each farmer is continuing to withdraw his share of the groundwater reserves. This is possible because each farmer's share of the reservoir water has been cut back proportionately,

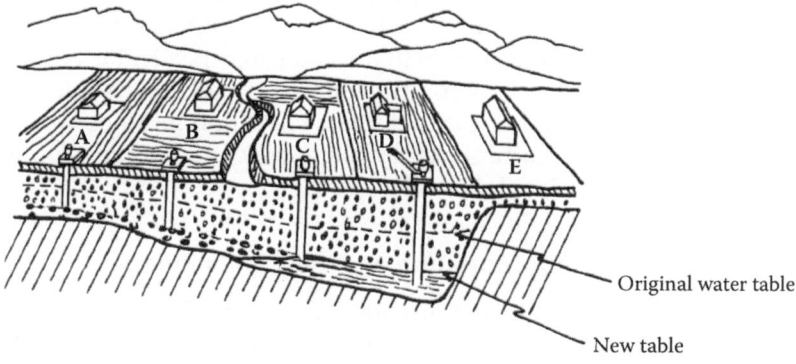

Original water table

New table

FIGURE 56

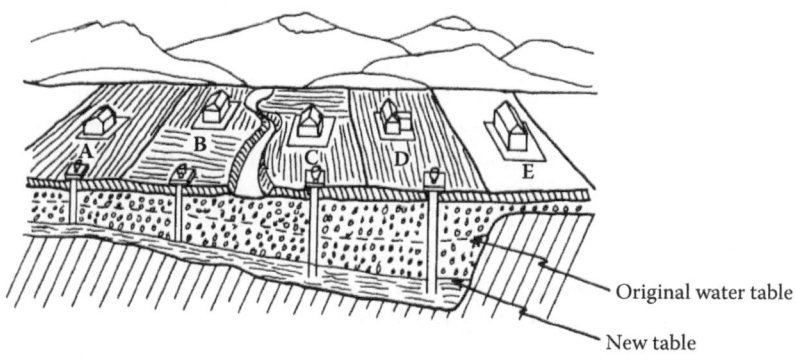

Original water table

New table

FIGURE 57

allowing each to continue withdrawing water from the reservoir, though not at the rate they are accustomed to. Thus, the correlative rights rule simply divides the remaining water in a drought-stricken reservoir in such a manner that each user has a share amounting to a total withdrawal that would not deplete the reservoir.

All three of the groundwater doctrines discussed in the preceding paragraphs are similar in one major aspect. They are all based on the direct ownership of land above the reservoir. Under these doctrines a water user must always own land overlying the reservoir in order to gain rights to the use of the water during a shortage. The correlative rule, however, is the only doctrine where the landowner does not actually own the water beneath his land but rather, he has only a right to its use, and the use must be beneficial and reasonable even when applied to his overlying land.

Should any individual gain ownership of a piece of land in a water basin controlled by any one of these three doctrines, he immediately has the right to sink a well and begin drawing water from the reservoir. This would apply no matter how many users were already in the basin. The new landowner's right to a share in the basin would apply even if drought conditions were in existence. Under the California rule this could cause particular hardship in small overdeveloped areas. Eventually each user might receive such a small share of the water in the reservoir that they would find it impossible to continue irrigated farming. In this country the question of overdevelopment is usually settled by the prior-rights rule.

The Doctrine of Prior Appropriation or Prior-Rights Rule

During the California gold rush of 1848, the miners established their own rule concerning water rights. The custom or law was simply first come first served, which meant that the first one to use a stream had the right to its continued use over all later users. This law gradually grew into the doctrine of prior appropriation or prior rights. And unlike the other three groundwater doctrines, it has been used to regulate the use of surface and groundwater alike.

> Under this rule no ownership rights are granted. The individual user only has the right to the use of the water. Each appropriation is limited to the beneficial needs of the land regardless of whether the ground overlies the groundwater reservoir, which means that it is perfectly legal to export the water from the basin as long as it is used for a beneficial or reasonable purpose. After a water right is established, however, it remains in effect only as long as the water use conforms with existing regulations.

> Under this doctrine the senior or first appropriator of water from an underground source has preferred rights. His right to the use of the water from the basin is protected over all junior or later appropriators

of water from the same basin. During water shortages, junior appro-priators can be compelled by law to shut off their pumps. The pumps are then shut off in the reverse order of the dates on which the wells were drilled.

The states which have adopted this system usually have some regu-latory agency charged with the responsibility of controlling over-development in groundwater basins. This usually involves the state engineer who has the authority to withhold drilling permits in areas of overdevelopment. And in most instances, the state engineer's deci-sions on withholding permits in overdeveloped areas are subject to court review.

Groundwater Legislation in Colorado

Surface water in Colorado is legislated by prior appropriation and the groundwater laws following were added:

Prior to 1957, no basic legislation existed in Colorado concerning a doctrine to regulate groundwater withdrawal. Then, on May 1, 1957, the Colorado state legislature enacted a groundwater law that provides control over "any water not visible on the surface of the ground under natural conditions."

The law provides that within 3 years from the effective date of act all groundwater users had to register existing wells with the state engineer. New wells cannot be drilled nor the production of water from existing wells increased unless the user applies to the state engineer for a "Permit to Use Groundwater." The state engineer will issue the permit unless it comes from within a "Tentatively Critical Groundwater District."

The law established a State Groundwater Commission composed of eight members appointed by the governor. The commission was then given the authority to conduct a preliminary survey and designate Tentatively Critical Groundwater Districts. All areas in the state in which it appeared that the withdrawal of groundwater had "approached, reached, or exceeded the nor-mal annual rate of replenishment" were to be declared tentatively critical. The law further provides that the commission may, at any time, or at the request of the state engineer, or upon petition of a substantial number of well owners within an area of the state, investigate and declare an area to be a Tentatively Critical Groundwater District.

After a district is declared tentatively critical, the commission establishes boundaries and closes the area to further groundwater development. The state engineer has the power to refuse to issue "Permits to Use Groundwater" in critical districts except for:

1. Wells used solely for stock watering purposes
2. Domestic wells having discharge pipes of 2 in. or less
3. Artesian wells with discharge pipes not exceeding 3 in. in diameter

4. The replacement, deepening, or reconstruction of wells in need of restoration which have been in operation for more than one year prior to the date the law took effect

The Colorado groundwater doctrine is basically a variation of the prior-rights rule. The state can regulate and control the amount of new well development in a critical area, but it does not regulate the use of the water in the basin by established well owners. It is therefore apparent that the Colorado law in its present form cannot prevent the total depletion of an underground reservoir

Some Legal Concepts and Definitions

NOTE: Listed below are some legal concepts and definitions relating to cadastral (land) surveying. The list is by no means complete and can serve at best only as an introduction to a very extensive field of study. As with all legal matters these concepts are to be used with CAUTION and a LAWYER SHOULD BE CONSULTED.

Role of Surveyor or Engineer

Within the context of land surveying, the role of the surveyor is very carefully defined by law.

Court or legal system—In the United States, land can only be transferred through legal action, and appropriation without due process is illegal. The right to own land is one of the fundamental rights stated in the Fifth Amendment to the U.S. Constitution.

REGISTERED surveyor—A person legally registered in a state to practice land surveying. A registered surveyor can be an expert witness to the court, who through his knowledge attests to physical facts, such as length, angles, elevations, etc., and mathematical calculations. Without court action, a surveyor CANNOT establish ownership of land.

Surveyor, engineer—A person not legally registered has no legal standing in the eyes of the law. As a consequence his work is of no legal value. Registration as an engineer normally does not establish surveying competency.

Covenant

A restriction written into a deed, restricting the use or occupancy of the land by its owner. It is binding on all subsequent purchasers. For example: only five horses per lot, no tar paper buildings, solar covenants may restrict building heights to prevent shadows on adjacent properties, etc. Especially in subdivisions covenants are very easy to establish, but nearly impossible to rescind, because 100% agreement is required.

Easement

A written permission by the landowner for a specific purpose. There are overhead, surface, and subsurface easements. For example: overhead power line, access road, sewer line. Commonly easements are granted to utility companies (electric, gas, water, sewer, cable TV, etc.) for a specific time period (99 years, etc.). The landowner retains the use of the land, but the utilities have the right-of-access. For example, bringing in a truck to fix overhead wires, or digging up a broken waterline, etc. Damage liability is normally limited. Construction can have a construction easement (maybe 50 ft wide) and a permanent easement (maybe 10 ft wide). An easement should be described by a metes and bounds description and then filed and recorded with the property deed in the courthouse.

Examples of an unusual easement are the irrigation ditch "easements" in Colorado, where the ditch easements are protected by the Constitution of the State, Article 2, Section 14. Whether or not the ditch easements are recorded or are historical, a ditch has the dominant right and the property owner does not have the right to interfere with the location and access for maintenance of an easement. In 2001 the State Supreme Court ruled (2001 WL 1456156), again, that:

1. Ditch easements may NOT be altered by the property owner, and
2. Changes may only be made which benefit the ditch owner, but NOT the property owner, i.e., the ditch company has to have access to the ditch, and can clean, widen, or even realign a ditch without compensation to the landowner.

Buyer beware: In essence, an irrigation ditch is more important than a property right, and the ditch easements are probably not recorded.

Transfer of Real Property (Real Estate)

In the United States real property (real estate) can only be transferred through legal action, and appropriation without due process is illegal. A surveyor can only attest to the physical facts, length, angles, etc., and mathematical calculations, but he CANNOT establish ownership.

A legal definition of real property is: "The interest that a man has in lands, tenements or inheritances, and also in such things as are permanent, fixed and immovable and which cannot be carried out of their places, as land

and tenements" (Brown, p. 84); another is: "land and generally whatever is erected, growing, or affixed to the land" (Madson. p. 787).

In general there are only two legal methods whereby real property can be transferred; they are through death (descent) or purchase, and the conveyance must be in WRITING, for example, a will or a written deed.

Deed

A deed is probably the most common form of conveying real property interests. Even though there are variations from state to state, the requisites or essentials of a deed are:

1. *Competent or proper parties.* Competent parties are persons who know the nature (content) of their action, generally a person of legal age. The proper parties are the grantor (seller) and the grantee (buyer). Either party can be represented under a power of attorney or by a lawyer.

2. *Proper subject matter.* The grantor must have tangible interest in the real property that is to be transferred. A deed is invalid when the grantor (seller) has no real property interest. Possibilities or mere possibilities of an interest are not grantable by deed.

3. *Valid consideration.* The payment of money; one dollar may be sufficient.

4. *Written or printed form.* The writing can be either on paper or parchment.

5. *Sufficient or legal words.* The more completely and concisely a deed is written, the better it is because a deed represents a permanent record that will exist longer than the parties involved, i.e., the deed will exist "forever." The description must contain the name of the grantor (seller) and the grantee (buyer), the complete description of the property boundary, easements, mineral rights, etc., consideration (money), the signatures, date, and seal(s).

6. *Reading before execution of deed.* The principal parties must have an opportunity to read the written deed before its execution.

7. *Execution, signing, sealing, attestation.* The principal parties must sign, and the signing must be witnessed and sealed by a notary.

8. *Delivery.* Delivery is the concept where the terms of the deed have been fulfilled. A manual delivery by the seller (grantor) or his agent is desirable but not required. The effective date of a deed is the date of delivery and not the date of signing, etc., and the deed then becomes a binding document for both grantor (seller) and grantee (buyer).

NOTE: The above steps complete the transactions of a deed; however, in most U.S. jurisdictions it is further required that the deed is filed by date and by record number in the local courthouse or parish. An unfiled deed is inferior to a filed deed.

Quitclaim

A written conveyance whereby the grantor (seller) conveys whatever interest he *may* have, without warranty of title. A quitclaim is a common and quick, but legally hazardous means of obtaining construction easements.

Adverse Possession

A legal method for acquiring title to land under certain conditions. There are considerable variations, but in general the following requirements are:

1. *Actual possession.* Can be by the claimant or by his tenant or agent. Payment of taxes is not sufficient to establish actual possession.
2. *Open and notorious.* The claimant must give notice to the "world" about his intentions such that the true owner cannot be deceived, i.e., he/she must be seen during daylight.
3. *Continuous.* The occupancy must be continuous during the statutory limit, generally 18 or 21 years. For example, the continuity requirement can be voided by vacating or closing a gate to the property for one day per year.
4. *Exclusive.* The claimant must be the exclusive user of the land.
5. *Hostile.* The claimant must be in possession as an "owner."

NOTE: In addition to the items listed above, color of title is of great evidentiary value in establishing adverse possession.

Chain of Title

The "chain of tile" is a continuous written and recorded record of ownership of a specific piece of property, telling who bought it and sold it, and when and what encumbrances are on the property, from the first owner (survey) to the present. In the western U.S. and major parts of Canada, the title record must start with the original GLO (General Land Office) survey; in the eastern U.S. and elsewhere it must start with the first metes and bounds survey. A "broken chain of title" or a "defective title" may make ownership after the break doubtful or will put a "cloud" on the title of the present owner. Due to fires and natural disasters, it may be impossible to obtain a complete chain of title from courthouses or other official sources. With a fully paid property, the chain of title documents should be with the landowner; with a mortgaged property the mortgage owner (bank, mortgage company, etc.) normally holds the documents.

A title abstract usually extracts only the pertinent information, often only partially, and it cannot be substituted for the actual chain of title documents filed, on paper and electronically, in the county or parish archives.

Title insurance, "a policy of insurance which indemnifies the holder for loss sustained by reason of a defect in the title, provided the loss does not result from a defect excluded by the policy provisions" (Colorado Real Estate Manual).

Relocation of Property from Previous Survey Work

In principle, the survey of land (Cadaster Survey) starts with large tracts of land which are, over time, broken down with new surveys and resurveys into ever smaller parcels. Each survey produces monuments and documentation, which are entered into the legal system. No matter when a property was first surveyed, many things can happen to the property and its records. It is therefore imperative that a priority is established by which a resurvey is documented. The order is from highest to lowest:

1. *Monument.* Of primary importance is the monument set in the original survey and called out in the original survey notes. For example: cross chiseled in bedrock, 3 × 4 × 1 ft granite boulder, car axle, rebar, red oak hub, pine stake, etc. Of secondary importance are witness comers, accessory monuments, or simply accessories, which may have been set as "backup" in case the original monument is destroyed. If properly done, it is permissible to replace an original monument with a new, "better" monument in EXACTLY the same location, and then file and record the "new" monument with the property deed. A resurvey, or "following in the steps of the original surveyor," will often show an "error" in the original survey, or improved accuracy due to more modem equipment. It makes absolutely no sense (legally or otherwise!) to set a "new" monument within a few feet of the original. It does make considerable sense to survey into the original monument and then accept the "new" measurement as the legally binding value. It is of questionable merit to claim a fraudulent survey when the monument

has been in place and accepted for years. Monuments are seldom lost, but more often are obliterated (i.e., covered with dirt, overgrown with vegetation, etc.), and a diligent search will probably recover 80–90% of all monuments.

2. *Written description.* The written description is inferior to the monument. In the absence of a monument, the legally filed and recorded written description controls. Descriptions which are not filed and recorded should be approached with GREAT CAUTION.

3. *Plat or labeled drawing.* A legally filed and recorded plat in turn is inferior to the written description. Often a comparison between the written description and the plat is helpful because one or the other may have typographical or transcription errors.

4. *Area.* Even though land is mostly sold by area, it often is impossible to determine the "exact" boundary line when there are no monuments or written descriptions. Adjacent property descriptions may be of some help. Land areas are always reported on the horizontal, and the area is given as "more or less."

5. *Testimony of an "old timer."* Especially in rural areas, longtime residents may remember the location of monuments or surveying activity. If possible, several witnesses should attest to the same fact (i.e., location of a monument).

Property Rights Abutting Water

Riparian Property Rights

Riparian rights pertain to property rights along a river or stream. Also depending upon the state, the riparian property owner may NOT have the *water rights* in the river, (i.e., the water does not belong to the property owner). In general two concepts are applicable: (1) property to the bank of a navigable river and (2) property to the centerline of a non-navigable river.

1. Property to the Bank of a Navigable River

Generally the bank of a river is considered to be the ground scoured by the moving water, and where the land is unsuitable for agricultural purposes. The river bottom between the banks is usually state property. In arid areas the bank is often considered to be the "mean" water line.

The term "navigable" is vague at best. In the past, the U.S. Supreme Court legally defined a navigable river which in "its ordinary condition, has been or can be used as a highway of commerce... in the customary modes of trade or travel." For example: steam riverboats, commercial rafting, log rafts, etc.

A navigable river can only be defined by court action and the U.S. Army Corps of Engineers is charged with defining a "navigable river" and the "point of navigability."

The definition above worked well with surveyors until the U.S. Supreme Court in *Rapanos v. United States,* 547 U.S. 715 (2006) redefined the term "navigable waters;" in turn on October 9, 2008, the 11th Circuit Court in Case No. 05-17019 wrote the definition that navigable waters are "streams which may eventually flow into a navigable stream or river." Both cases resulted from an effort to allow the U.S. Corps of Engineers to exercise federal control over rivers and wetlands in order to release money from the Clean Water Act. Undoubtedly there will be other subsequent lawsuits clarifying these rulings with respect to property law and surveying practices.

With all navigable rivers the "point of navigability" must be considered (i.e., the highest point along the course of the river where it becomes navigable, normally a bridge or a prominent landmark).

NOTE: In many states rivers, streams and lakes and more recently some beaches have been legally ruled to be public domain, thus the concept of a navigable river is not applicable.

2. Property to the Center of a Non-Navigable River

Above the point of navigability the property line is the thread (midpoint between opposite banks) of the river or stream.

Littoral Property Rights

These property rights pertain to the shore, especially the seashore:

> **Upland rights.** In most states the upland private littoral landowner has title above the mean high-tide mark. In Maine, New Hampshire, and Massachusetts, the private littoral rights include the tidal flats (i.e. down to the mean low-tide mark); hence, these laws prevent the public from using the beaches. Both along the east and west coasts there have been and are legal challenges to allow the public access to all beaches.
>
> **Tidelands.** The land between the mean-high and mean-low tide (i.e., the beach) is public domain and belongs to the state, which may dispose of it. The exceptions are Maine, New Hampshire, and Massachusetts, as well as some portions on the west coast. The "modern" trend in jurisprudence tends to make the beaches public domain land.
>
> **Offshore rights.** On May 22, 1953, a U.S. Congressional bill granted the states a 3-geographical mile (1 geographical mile = 6,076.10333 ft)

seaward limit, measured from the mean low-tide line. Exceptions are in the Gulf of Mexico from Key West, Florida, westward along the coast to the Mexican border at Brownsville, Texas, where a 9-geographical mile limit exists. The states acquired fee title, mineral rights (primarily oil, natural gas, sulfur, and salt), but the federal government reserved the rights to control commerce and navigation, etc.

Beyond the 3 (or 9) geographical mile limit. In this case, the U.S. federal government claims title (mineral rights, sunken treasures, etc.) seaward to the continental shelf. The continental shelf is defined as a gentle sloping underwater plain, extending outward into the ocean, generally less than 500 ft deep and terminated by the steeper continental slope.

Land gain by accretion or loss by erosion or inundation. In most states the riparian or littoral owner gains or loses, when the change is due to *natural causes*. Inundated land reverts to the owner of the river, lake, or ocean bed—probably the state. A naturally formed island in the river or lake belongs to the owner of the bed, again probably the state. Property lines remain as determined prior to artificial or manmade accretion or as a result of man's activities of shore modification with levies or breakwaters, etc.

Registry of Property—The Torrens System

In an abstract format with the Torrens system, property is registered with the name of the owner; with the "conventional system," property is registered by metes and bounds and/or by aliquot parts of the GLO survey.

In principle, there are two fundamentally different ways to register the ownership of land, (i.e., file and record property):

1. **By deed recordation:** The description of the boundary is by metes and bounds. The primary reference is the boundary description and the secondary is the owner's name. Each deed relies on the previous deed and each survey relies on the previous survey, thus for a valid title there must be an unbroken chain of title and surveys without faults.

2. **By title recordation:** The recording of the property is by the owner's name (Torrens Title). The primary reference is the owner's name and the secondary is the property description. Each registered title is conclusive proof of ownership.

The Torrens System

The Torrens system is named after Sir Richard Torrens (1814–1884) who developed the concept in 1858 for New South Wales (Australia). It is in current use in Australia, parts of the British Commonwealth, and partially in Minnesota, Massachusetts, and the city of Chicago. The Torrens system was made permissive in Colorado in 1903, and it still is in the state law as an alternate method of land title registration (CRS 38-36).

The Torrens act has three basic aspects:

1. A judicial determination of title is made, which is similar to a quiet title action and then becomes absolute against all parties. A governmental entity (state, county, city, etc.) operates the system and guarantees the title to be free of encumbrances.
2. The governmental land office registers and certifies all future transactions against the title, without any additional title search.
3. An assurance fund provides compensation to a person, who through no fault of his/her own, has suffered as a result of the initial judicial quiet title action. This fund is funded by land registration fees.

Although the Torrens system will certify title, it does not eliminate boundary disputes of the registered land. It only isolates and probably simplifies the issues by reducing the dispute to a single property line rather than involving all parties bordering the entire property and its ties.

A major deterrent for the adoption of the Torrens system is that the state (court) must guarantee the right of ownership and pay for any error in the initial determination of who has the right of ownership.

Glossary

Every profession has its own vocabulary and land surveying is no exception. Given below is a broad and not too technical description of abbreviations and terms or concepts relating to property surveying practices. For a more comprehensive list and standard definitions the reader is referred to *Definitions of Surveying and Associated Terms*, by ACSM, referenced in the Bibliography.

Abbreviations

BLM—Bureau of Land Management: In 1812 Congress established the General Land Office (GLO; see below) in the Department of the Treasury; in 1946 the Bureau was housed within the Department of the Interior. The agency is responsible for managing federal lands, about 264 million acres (roughly 4112,500 sq. mi.) and their natural resources.

EDM—electronic distance meter: instrument to measure distances electronically; predecessor to modern Total Station (see below), which can be used to measure distances and angles.

GIS—Geographic Information System: computer-based databases for maps, cadastre files, etc.

GLO—General Land Office: originally the federal agency responsible for all national land surveys. In 1946 it became BLM.

GLONASS—Global'naya Navigatsionnaya Sputnikovaya Sistema: Russian-made satellite system; partially operational.

GPS—Global Positioning System: U.S.-made satellite system, became fully operational in 1994, uses 24 satellites.

HARN'92—High Accuracy (horizontal) Reference Network 1992: superseded by NSRS'06.

NGS—National Geodetic Survey: the federal agency which established and maintains the national horizontal and vertical control network.

NAD'27—North American Datum 1927: National horizontal reference system, 1927, uses Clarke 1866 ellipsoid; superseded by NSRS'06. The original data set for horizontal control stations across the United States; published until about the mid-1980s.

NAD'83—North American Datum 1983: National horizontal reference system, 1983, uses GRS'80 ellipsoid. Data set representing the first general adjustment of about 200,000 horizontal control stations across the United States. Updated to NSRS'06.

NAVD'88—North American Vertical Datum of 1988: the "new" elevation datum for Canada, the United States, and Mexico.

NGVD'29—National Geodetic Vertical Datum 1929: the original "mean sea level" shown on most topographic maps, superseded by NAVD'88.

NSRS'06—National Spatial Reference System 2006: Data set representing the most recent general adjustment of about 270,000 horizontal control stations across the United States.

P.O.B.—Point of beginning: starting point of a property survey.

USGS—United States Geological Survey: A branch of the Department of the Interior. Its mission: "to provide reliable scientific information to describe and understand the earth." A valuable resource of maps and aerial photos.

UTM—Universal Transversal Mercator coordinates.

Terms and Concepts

Area: The amount of land enclosed by a property survey. Either originally surveyed horizontally or later converted mathematically to a horizontal surface.

> **Acre:** one acre = 43,560 sq. ft.
>
> **Square m.:** one square meter-about 10.76 sq. ft.

Azimuth: Direction of a line based on angular measurements turned clockwise from true north, maximum value = 360°. See Figure 4.

Bearing: The horizontal direction of a line in degrees, minutes, and seconds of arc, east or west from a true (or magnetic) north or south direction. Maximum value = 90°. See Figure 3.

Chain: Originally a Gunter chain, 66 ft long, 100 links, used for surveying distances. In modern vernacular, a chain is a 100-ft steel tape allowing measurements to the nearest 1/100 ft. The fundamental unit for the GLO rectangular survey system. Even though all distances are stated in chains and links they can be, and usually are, converted to decimal feet.

Chaining: Modern vernacular for making "horizontal" measurements with a steel tape.

Chain of title: A continuous written record of property titles for a specific parcel of land. In the GLO states the title must start with the initial conveyance of the public land to an individual. A "broken chain of title" refers to a discontinuous record. Sometimes title records may be obtained from title companies; due to fires, floods, hurricanes, etc., it may be impossible to obtain a complete record. See "File and Record" below.

Call to monument (also just "call"): The distance and direction (bearing or azimuth) to a monument.

Decimal feet: Surveyors use a (survey) foot subdivided into decimals, i.e., 1/10 ft and 1/100 ft. A 0.01 ft is roughly 1/8 in. Surveyors do not use inches!

Ellipsoid: Mathematical model of earth, a solid generated by rotating an ellipse.

    Clarke 1866: Ellipsoid specified by British geodesist A. R. Clarke in 1866. Used for NAD'27 data base. GRS'80: Geodetic Reference System 1980, the ellipsoid used by GPS and NAD'83/NSRS'06 data base.

Elevations (also called "heights"): The vertical distance from either mean sea level or the geoid.

    1. Orthometric elevations: Elevations determined conventionally with a spirit level and referenced to mean sea level.

    NGVD'29: National Geodetic Vertical Datum of 1929; no longer supported by NGS, the elevations shown on most maps and benchmarks.

    NAVD'88: North American Vertical Datum of 1988; the "new" elevations, about 4 to 5 ft. higher than NGVD'29 elevations.

    2. Ellipsoid elevations (heights): Elevations referenced to a specific ellipsoid, i.e., GPS elevations. Can be different from orthometric elevations by as much as 100 ft.

File and record: The process of entering a document into the legal system at the courthouse. Upon receipt of either a paper or electronic document, the county clerk and recorder time-stamps the document with date and time in hours, minutes, and seconds and a record number. Subsequently all documents are (were) collated sequentially by time and ultimately bound into volumes of up to 500 pages each. A title search conceptually starts at Vol. 1 and ends at the last entry of the current day. See "Title Search" and "Chain of Title."

Foot: The United States is the only country in the world where two basic units of length are legally used:

    1. The U.S. Standard Foot: also called the International Foot, is used for all measurements, except

    2. The U.S. Survey Foot: is used for both plane and geodetic surveys. It is the derived unit used in Total Stations, GPS, State Plane and UTM Coordinates, NGS National Geodetic data base, mapping, etc.

    Foot (U.S. Standard): usually subdivided into 12 inches.

    The U.S. adopted the metre as the primary standard of length with: U.S. Coast and Geodetic Survey Bulletin 26, by Thomas C. Mendenhall, April 5, 1893.

On July 1, 1959 the U.S. Standard foot was redefined as:
1 foot (U.S. Standard) = 0.3048 metre exactly
Foot (U.S. Survey), usually subdivided into 1/10 and 1/100 ft. By US Coast and Geodetic Survey Bulletin 26, April 5, 1893 the foot was defined as:
1 foot (U.S. Survey) = 12.00 in./39.37 in./m
= 0.304 800 609 metre
Note: all geodetic and surveying measurements in the U.S. remain based on the original (1893) definition of the foot and were not changed in 1959.
Practical consideration: Since the survey foot is about 2 ppm (parts per million) longer than the standard foot, a 5,000 ft(survey) line would be about 1/100 longer than measured with a standard foot, therefore for any measurement over 1 mile the difference is significant.

Geoid: Theoretical shape of "sea level" extended through the continents, perpendicular to gravity.

GLO states: States having the General Land Office (GLO) survey, i.e., the rectangular government survey.

Gunter's chain: Surveying measuring chain, invented by Edmund Gunter about 1620; 66 ft. long, 100 links, made of iron.

Lieu land, mostly as in "lieu section": Usually associated with railroad land grants where the RR received or traded land in "lieu of" the land from the original grant; mostly due to prior occupation by private land owners.

Mapping angle (also called "grid angle"): The angle between true north and coordinate or grid north. Can be positive or negative and is projection dependent.

Metes and bounds: A consecutive verbal description for the boundary of a property by distance and direction of each leg; can have curves. Must close back to point of beginning (POB).

Meridian: A true north–south line on the surface of the earth, a great circle route.

Monument (or "surveying monument"): A surveying marker; historically made from anything handy, such as a stone stood on end, steel pipe, rebar stood on end, etc. A modern monument can be a brass or aluminum cap set in concrete and engraved with the surveyor's license number.

Monument of call: A surveying monument described in the survey notes, preferably the original survey.

National Horizontal Control Network: A network of about 270,000 control monuments located throughout the United States. Established and maintained by NGS.

National Vertical Control Network: A network of about 750,000 control monuments (benchmarks) located throughout the United States. Established and maintained by NGS.

North

    True north: Referenced to geographic north, usually determined astronomically.

    Magnetic north: Referenced to magnetic north, determined with magnetic compass, varies with time and place of observation, deviates from true north by magnetic declination.

    Grid north: Alignment of north at the center of a map, with used plane coordinates and GPS.

    Magnetic Declination (also called "variation"): Deviation between magnetic and true north, varies with place and time.

Orthometric elevations (sometimes called "heights"): Ground elevations carried across the United States with conventional levels. Differ by geoid height from ellipsoid elevations obtained with GPS.

Plat (map): A drawing showing all surveying information for a land property, usually a subdivision.

Principal point: The starting point for a GLO survey, for example: 6th Principal Meridian.

Projections: The mathematical principles of "flattening" the curvature of the earth into a plane.

    Lambert system projection: Named after J.H. Lambert, the surface of the earth is projected onto a cone.

    Mercator projection: Named after Gerardus Mercator, the surface of the earth is projected onto a cylinder.

Random line: A surveying method to establish a straight line, when the endpoints are not intervisible. Starting at one endpoint, a "random" line is run approximately in the direction of the other endpoint. Once the second endpoint is sighted, angles and distances will allow the true line to be established.

Range box: A metal or plastic collar covered with a lid, set on grade, i.e., like a miniature sewer manhole. The range box is used to protect the monument set in the inside.

Registered (licensed) surveyor: A person legally registered in a specific state to practice "land" surveying in that state only; can be an expert witness to the court. Without court action a surveyor *cannot* establish a property line. A person not legally registered has no legal standing in the eyes of the court.

Section: A subdivision within a township of the public land, containing 1 square mile more or less, or 640 acres more or less.

    Lieu section: A section of land which was acquired "in lieu of" or "in place of" another section. The grantee had the right to refuse grant land and request another section. Usually applied to land grants by the United States to railroads, canals, colleges, etc.

Side shots: A field surveying method, where multiple measurements are made in all directions from the same instrument setup. Usually not desirable, because there are no physical or mathematical checks.

State plane coordinates: A plane rectangular coordinate system used by surveyors or engineers allowing them to reference ground features, property lines, etc., in proper proportions to each other. Unlike property lines tied to a monument; once established, items referenced by state plane coordinates can be relocated from any known point. Can be expanded to include 3-D coordinates.

Great/small circle: On the spherical earth a great circle has its center at the center of the earth, i.e., the equator, a meridian passing through the poles; a small circle can be a line of equal latitude.

Subdivision by aliquot parts: Subdividing the public land into (integer) parts without leaving a partial or fractional remainder.

Ties: Distance and direction (bearing or azimuth) from a monument to a reference monument.

Title search: A search of the written records, usually in a courthouse, pertaining to a specific property. Also see "chain of title" and "file and record," shown above.

Total station: A modern surveying instrument capable of electronically measuring distance and angles, usually requires a reflector.

Township: A subdivision of the public land, containing 36 square miles more or less.

Township (T)/Range (R): The "coordinate" designation of a township in the GLO survey system, for example: Section 4, T4S, R70W, Crow Meridian.

Traverse: There are two types of traverses: an open and a closed traverse.

> **Open traverse:** A series of legs defined by distance and angles (direction) that physically do not return to the point of beginning. Can also apply to an open elevation (level) traverse or open level loop.

> **Closed traverse:** A series of legs defined by distance and angles (direction) that physically return to the point of beginning. Can also apply to a closed elevation (level) traverse or closed level loop.

Variation, abbr. "var." (also called magnetic declination): Deviation between magnetic and true north; varies with place and time.

Bibliography and Other Resources

Bibliography

The following list of references includes current texts and classics now out of print. Readers should keep in mind that new editions appear regularly and that publisher data, Internet sites, and email addresses are subject to change. Recent surveying texts may not cover GPS completely and many GPS textbooks are not amenable to self-study. Finally, even older texts will contain helpful information. Out-of-print texts may be available from Amazon.com or other used book sources.

Allen, Arthur L. 2007; *Principles of Geospatial Surveying*. CRC Press, Boca Raton, FL.

American Congress on Surveying and Mapping and American Society of Civil Engineers. 2005; *Definitions of Surveying and Associated Terms*. Gaithersburg, MD. acsm.net

Anderson, James M., and Mikhail, Edward M. 1997; *Surveying: Theory and Practice*. McGraw-Hill, New York. mhhe.com

Bartlett, Richard A. 1962; *Great Surveys of the American West*. University of Oklahoma Press, Norman.

Beck, Warren A., and Hasse, Y.D. 1989; *Historical Atlas of the American West*. University of Oklahoma Press, Norman.

Black, Henry Campbell, *Black's Law Dictionary*, 8th Ed., Garner, Bryan A., Ed., West Publishing Company, Eagan, MI.

Breed, Charles B., Barry, Austin B., and Bone, Alexander J. 1971; *Surveying*, 33rd Ed. John Wiley & Sons, New York.

Brown, Curtis M., and Eldridge, Winfield. H. 1962; *Evidence and Procedure for Boundary Location*. John Wiley & Sons, New York.

Brown, Curtis M., Landgraf, F.H., and Uzes, F.D. 1969; *Boundary Control and Legal Principles*. John Wiley & Sons, New York.

Brown, Curtis M., Robillard, W.G., and Wilson, D.A. 1995; *Brown's Boundary Control and Legal Principles*, 4th Ed. John Wiley & Sons, New York.

Brown, Lloyd A. 1949; *The Story of Maps*. Bonanza Books, New York.

Buckner, R. Ben. 1983; *Surveying Measurements and their Analysis*. Landmark Enterprises, Rancho Cordova, CA.

Buckner, R. Ben. 2001; *Land Survey Review Manual*. CRC Press, Boca Raton, FL

Clawson, Marion. 1968; *The Land System of the United States: An Introduction to the History and Practice of Lane Use and Lane Tenure*. Beard Books, Frederick, MD.

Code of Federal Regulations, Title 43: Public Lands: Interior, Part 3832. Available electronically: http://ecfr.gpoaccess.gov

Colorado State Board of Registration for Professional Engineers and Professional Land Surveyors. *Laws of the State of Colorado Regulating the Practice of Land Surveying.* Department of Regulatory Agencies, Denver.

Colorado State Board of Registration for Professional Engineers and Professional Land Surveyors. *Policies of the State Board of Licensure for Professional Engineers and Professional Land Surveyors.* Department of Regulatory Agencies, Denver.

Davis, Raymond E., Anderson, James, Foote, Francis S., et al. 1997, *Surveying Theory and Practice,* 6th Ed. Land Surveyors Publications, Spruce Pine, NC.

Egles, Yves, and Kasser, Michael. 2002; *Digital Photogrammetry.* CRC Press, Boca Raton, FL.

Elgin, Knowles, and Senne. *Celestial Observation Handbook and Ephemeris.* Sokkia Corporation, Olathe, SK. sokkia.com

El-Rabbay, Ahmed. 2006; *Introduction to GPS: The Global Positioning System.* Artech House, Norwood, MA. artechhouse.com.

Garner, Bryan A., Ed. 2004; *Black's Law Dictionary,* 8th Ed. West Publishing, Eagan, MN. west.thomson.com

Ghilani, Charles D., and Wolf, Paul R. 2006; *Adjustment Computations: Spatial Data Analysis,* 4th Ed. John Wiley & Sons, New York.

Holbrook, Stewart H. 1948; *The Story of American Railroads.* Crown Publishers, New York.

Kaula, William M. 2000; *Theory of Satellite Geodesy: Applications of Satellites to Geodesy.* Dover Publications, Mineola, NY. dover.com

Leick, Alfred. 2004; *GPS Satellite Surveying,* 3rd Ed. John Wiley & Sons, New York.

Madson, T.S., II, and Seemann, Louis N.A. 1980; *Fading Footsteps.* Land Surveyor's Seminars, Gainesville, FL.

McCormac, Jack C., and Anderson, Wayne. 1999; *Surveying.* John Wiley & Sons, New York.

Mikhail, Edward M., Bethel, James S., and McGlone, J. Chris. 2001; *Introduction to Modern Photogrammetry.* John Wiley & Sons, New York.

Mikhail, Edward M., and Gracie, Gordon. 1981; *Analysis and Adjustment of Survey Measurements.* Van Nostrand Reinhold, New York.

Minnick, Roy. 1980; *A Collection of Original Instructions to Surveyors of the Public Lands.* Landmark Enterprises, Rancho Cordova, CA.

Moffit, Frances H., and Bossler, John D. 1998; *Surveying,* 10th Ed. Addison Wesley Longman, Reading, MA.

Moffit, Francis H., and Bouchard, Harry. 1992; *Surveying,* 9th Ed. Harper-Collins, New York.

Noel, Thomas J., Mahoney, P.F., and Stevens, R.E. 1994; *Historical Atlas of Colorado.* University of Oklahoma Press, Norman.

Peters, William E. 1930; *Ohio Lands and Their History.* Reprinted 1979, Arno Press, New York.

Professional Land Surveyors of Colorado. 1997; *Colorado Revised Statutes Pertaining to Land Surveying,* Conifer, CO.

Robillard, Walter G., and Bouman, Lane J. 1987; *Law of Surveying and Boundaries,* 5th Ed. Michie Co., Charlottesville, VA.

Robillard, Walter G., Brown, Curtis M., and Wilson, Donald A. 2006; *Evidence and Procedures for Boundary Location.* 5th Ed. John Wiley & Sons, New York.

Robillard, Walter G., Wilson, Donald A., Wilson, Donald S., and Brown, Curtis M. 2003; *Boundary Control and Legal Principles.* 5th Ed. John Wiley & Sons, New York.

Seeber, Gunter. 2003; *Satellite Geodesy*, 2nd Ed. Walter de Gruyter, New York. degruyter. com

Sherman, Christopher E. 1925, *Original Ohio Land Subdivision.* Ohio Topographic Survey, Columbus.

Smith, James R. 1997; *Introduction to Geodesy: The History and Concepts of Modern Geodesy.* John Wiley & Sons, New York.

Stewart, Lowell O. 1935; *Public Land Surveys.* Collegiate Press, Ames, IA.

Stewart, Lowell O. 1935; *Public Land Surveys: History, Instructions, and Methods.* Reprinted 1976, Meyer Printing, Minneapolis, MN. info@meyers.com

Tillotson, Ira M. 1973; *Legal Principles of Property Boundary Location on the Ground in the Public Land Survey States.* Ira M. Tillotson, Missoula, MT.

U.S. Code Annotated, Title 43, Public Lands, Sections 1–1383. West Publishing Company, Eagan, MN. west.thomson.com

U.S. Department of the Interior, Bureau of Land Management. Date unknown; *Staking a Mining Claim on Federal Lands.*

U.S. Department of the Interior, Bureau of Land Management. 1973; *Manual of Instructions for the Survey of the Public Lands of the United States,* Technical Bulletin. 6. Also published 1855, 1881, 1890, 1894, 1902, 1919, 1947.

U.S. Department of the Interior, Bureau of Land Management. 1975; *Restoration of Lost or Obliterated Corners and Subdivision of Sections: A Guide for Surveyors.*

U.S. Department of the Interior, Bureau of Land Management. 1980; *Mineral Survey Procedure Guide.* GPO 857-030.

U.S. Department of the Interior, Bureau of Land Management. 1996; *Mining Claims and Sites on Federal Lands.* GPO 776-268.

U.S. Department of the Interior, Bureau of Land Management. 2007; *Locating Mining Claims: Information Guide.*

Van Sickle, Jan. 1997; *1001 Solved Surveying Fundamentals and Problems,* 2nd Ed. Professional Publications, Belmont, CA. goliath.ecnet.com

Van Sickle, Jan. 2001; *GPS for Land Surveyors.* CRC press, Boca Raton, FL.

Wattles, Gurdon H. 1976; *Writing Legal Descriptions.* Land Surveyors Publications, Spruce Pine, NC.

Wattles, Gurdon H. 1981; *Survey Drafting.* Land Surveyors Publications, Spruce Pine, NC.

Wattles, Gurdon H., and Wattles, William C. 1974; *Land Survey Descriptions.* Land Surveyors Publications, Spruce Pine, NC.

Wheeler, Keith. 1973; *The Railroaders.* Time-Life Books, New York.

White, C. Albert. 1983; *A History of the Rectangular Survey System.* U.S. Government Printing Office, 024-011-00150-6.

White, C. Albert. 1996; *Initial Points of the Rectangular Survey System.* Professional Land Surveyors of Colorado, Conifer, CO.

White, Richard. 1991; *A History of the American West.* University of Oklahoma Press, Norman.

Wilson, Donald A. 2007; *Forensic Procedures for Boundary and Title Investigations.* John Wiley & Sons, New York.

Wolf, Paul R., and DeWitt, Bon A. 2000; *Elements of Photogrammetry with Application in GIS.* 3rd Ed. McGraw-Hill, New York. mhhe.com

Wyckoff, William. 1999; *Creating Colorado.* Yale University Press, New Haven, CT.

Xu, Guochang. 2003; *GPS.* Springer Verlag, New York. springer-ny.com

Xu, Guochang. 2007; *GPS Algorithms and Applications* Springer Verlag, New York. springer-ny.com

Xu, Guochang. 2007; *GPS Software Library,* Springer Verlag, New York. springer-ny. com

Yang, Qihe, Snyder, John, and Tobler, Wald. 1999; *Map Projection Transformation: Principles and Applications.* CRC Press, Boca Raton, FL.

Other Resources

American Congress on Surveying and Mapping, 6 Montgomery Village Avenue, Suite 403, Gaithersburg, MD 20879. Phone: 240-632-9716. acsm.net

American Society of Civil Engineers, 1801 Alexander Bell Drive, Reston, VA 20191. Phone: 800-548-2723. asce.org

BNP Media, 2401 West Big Beaver Road, Suite 700, Troy, MI 48084. Free magazine subscriptions for surveyors. bnpmedia.com

Cheves Media LLC, *American Surveyor,* 905 West 7th Street, Suite 331, Frederick, MD 21701. Free magazine subscriptions for surveyors. amerisuv.com

Colorado State Board of Registration for Professional Engineers and Land Surveyors, 1560 Broadway, Suite 1300, Denver, CO 80202. Phone: 303-894-7788. aes@dora. state.co.us (email). dora.state.co.us/aes/index.htm# Information on applications and state portions of exams.

Land Surveyor Publications, 420 Oak Avenue, Spruce Pine, NC 28777. Phone: 800-533-4387. landsurveyors.com

National Council of Examiners for Engineering and Surveying, P.O. Box 1686, Clemson, SC 29633. Phone: 800-250-3196. Information on NCEES portions of examinations. ncees.com.

Professional Land Surveyors of Colorado, P.O. Box 704, Conifer, CO 80433. Phone: 303-838-7577. Information on membership and services. plsc.net.

Reed Business Geo Inc., 100 Tuscany Drive, Suite B-1, Frederick, MD 21702. Free magazine subscriptions for surveyors. profsurv.com

State Board of Registration for Professional Engineers and Professional Land Surveyors, Charles H. Adams, Program Director, 1560 Broadway, Suite 1300, Denver, CO 80202. Phone: 303-894-7788, Fax: 303-894-7790. E-mail: aes@dora.state.co.us. Web site: dora.state.co.us/aes/index.htm#. Information on applications and State portion of exams. Colorado Revised Statutes (Laws/Rules/Policies link—Colorado State Manager) http://198.187.128.12/Colorado/lpext.dll?f=templates&fn=fs-main.htm&2.0.

Taylor & Francis/CRC Press, 6000 Broken Sound Parkway NW, Suite 300, Boca Raton, FL 33487. Phone: 800-272-7737. crcpress.com

U.S. Naval Observatory, *The Astronomical Almanac.* Sold by Willmann-Bell Inc. P.O. Box 35025, Richmond VA 23235. Phone: 800-825-7827. willbell.com

World, 201 Sandpointe Avenue, Suite 500, Santa Ana, CA 92707. Free magazine subscriptions for surveyors. gpsworld.com

Journals

American Surveyor, Cheves Media LLC, 905 W 7th Street, Suite 331, Frederick, MD 21701. Free magazine subscription for surveyors. E-mail: amerisuv.com.

GPS World, 201 Sandpointe Ave, Suite 500, Santa Ana, CA 92707. Free magazine subscription for surveyors. E-mail: gpsworld.com.

Professional Surveyor, Reed Business Geo Inc., 100 Tuscanny Dr. Suite B-1, Frederick, MD 21702-5958. Free magazine subscription for surveyors. E-mail: profsurv.com.

P.O.B., BNP Media, 2401 W. Big Beaver Rd. Suite 700, Troy, MI 48084. Free magazine subscription for surveyors. E-mail: bnpmedia.com.

Index